AF358682

Energy Security as a Key Driving Factor for Socioeconomic Development: From Mitigation to Solution

Energy Security as a Key Driving Factor for Socioeconomic Development: From Mitigation to Solution

Editors

Giuseppe T. Cirella
Barbara Pawlowska

MDPI • Basel • Beijing • Wuhan • Barcelona • Belgrade • Manchester • Tokyo • Cluj • Tianjin

Editors
Giuseppe T. Cirella
Faculty of Economics
University of Gdansk
Sopot
Poland

Barbara Pawlowska
Faculty of Economics
University of Gdansk
Sopot
Poland

Editorial Office
MDPI
St. Alban-Anlage 66
4052 Basel, Switzerland

This is a reprint of articles from the Special Issue published online in the open access journal *Energies* (ISSN 1996-1073) (available at: www.mdpi.com/journal/energies/special_issues/energy_security_ key_driving_factor_socioeconomic_development_mitigation_solution).

For citation purposes, cite each article independently as indicated on the article page online and as indicated below:

LastName, A.A.; LastName, B.B.; LastName, C.C. Article Title. *Journal Name* **Year**, *Volume Number*, Page Range.

ISBN 978-3-0365-2689-8 (Hbk)
ISBN 978-3-0365-2688-1 (PDF)

Contents

About the Editors vii

Preface to "Energy Security as a Key Driving Factor for Socioeconomic Development: From
Mitigation to Solution" ix

Giuseppe T. Cirella and Barbara Pawłowska
Advancements in the Energy Sector and the Socioeconomic Development Nexus
Reprinted from: *Energies* **2021**, *14*, 8078, doi:10.3390/en14238078 . . . . **1**

Jennifer Hoody, Anya Galli Robertson, Sarah Richard, Claire Frankowski, Kevin Hallinan,
Ciara Owens and Bob Pohl
A Review of Behavioral Energy Reduction Programs and Implementation of a Pilot Peer-to-Peer
Led Behavioral Energy Reduction Program for a Low-Income Neighborhood
Reprinted from: *Energies* **2021**, *14*, 4635, doi:10.3390/en14154635 . . . . **7**

Olga Janikowska and Joanna Kulczycka
Just Transition as a Tool for Preventing Energy Poverty among Women in Mining Areas—A
Case Study of the Silesia Region, Poland
Reprinted from: *Energies* **2021**, *14*, 3372, doi:10.3390/en14123372 . . . . **35**

Patrick Rausch and Michał Suchanek
Socioeconomic Factors Influencing the Prosumer's Investment Decision on Solar Power
Reprinted from: *Energies* **2021**, *14*, 7154, doi:10.3390/en14217154 . . . . **49**

Ernest Czermański, Giuseppe T. Cirella, Aneta Oniszczuk-Jastrzabek, Barbara Pawłowska
and Theo Notteboom
An Energy Consumption Approach to Estimate Air Emission Reductions in Container Shipping
Reprinted from: *Energies* **2021**, *14*, 278, doi:10.3390/en14020278 . . . . **59**

Yu Fu, Agus Supriyadi, Tao Wang, Luwei Wang and Giuseppe T. Cirella
Effects of Regional Innovation Capability on the Green Technology Efficiency of China's
Manufacturing Industry: Evidence from Listed Companies
Reprinted from: *Energies* **2020**, *13*, 5467, doi:10.3390/en13205467 . . . . **77**

Giuseppe T. Cirella, Alessio Russo, Federico Benassi, Ernest Czermański, Anatoliy G.
Goncharuk and Aneta Oniszczuk-Jastrzabek
Energy Re-Shift for an Urbanizing World
Reprinted from: *Energies* **2021**, *14*, 5516, doi:10.3390/en14175516 . . . . **99**

About the Editors

Giuseppe T. Cirella

Giuseppe T. Cirella is a professor of Human Geography at the Faculty of Economics, University of Gdansk, Sopot, Poland, where he recently received his equivalency Doctor of Habilitation (Dr. Hab.) in Economics and Finance. He specializes in economic development, environmental social science, and sustainability. After completing a Doctor of Philosophy (Ph.D.) at Griffith University, Gold Coast, Australia, within the Centre for Infrastructure Engineering and Management, he founded the Polo Centre of Sustainability in Italy. He is also on the editorial board of a number of internationally-renowned academic peer-reviewed journals.

Barbara Pawlowska

Barbara Pawłowska is a habilitated doctor in Economics, a professor at the Faculty of Economics, University of Gdansk, Sopot, Poland. Her areas of specialization and research interests are sustainable development especially in the transport sector, environmental impact assessment, external costs of transport and energy estimations, and urban development. She is the co-author of a handbook prepared at the request of the European Commission on methods of estimating external transport costs for 28 EU member states (2009). She is also a member of the ministerial group for sustainable development and corporate social responsibility at the Ministry of Regional Funding and Policy. Since October 2019, she is the director of Doctoral School of Humanities and Social Sciences of the University of Gdansk.

Preface to "Energy Security as a Key Driving Factor for Socioeconomic Development: From Mitigation to Solution"

This book looks at the increasing demand for energy of contemporary societies and economies from around the world. Energy is the driving force behind development. As such, the future challenge will be not only to meet the rising demand, but also to implement less reliance on depleting fossil fuels, which cause damage to the environment. Moreover, the sustainability of supplied energy requires a reduction of emissions to control the absorption capacity vis-à-vis the environment. Globally, policymakers have largely recognized the significance of the relationship between energy and economic progress. Policymakers usually consider the social and economic aspects of energy security in terms of affordability and accessibility of service. The conditions of socioeconomic development depend on safe, secure, and sustainable energy at affordable prices. One of the prime concerns of policymakers should be to ensure energy security at the national level. These factors result in an increasing interest in undertaking activities in developing renewable resources. Energy efficiency is treated as the most cost-effective way to reduce energy demand while maintaining stable economic activity. Increasing energy efficiency is an important contributive aspect to solving issues in relation to climate change, energy security, and energy competitiveness. As a result, no country can afford to waste energy—giving rise to this Special Issue of "Energy Security as a Key Driving Factor for Socioeconomic Development: From Mitigation to Solution" in the journal *Energies*.

Giuseppe T. Cirella, Barbara Pawlowska
Editors

Editorial

Advancements in the Energy Sector and the Socioeconomic Development Nexus

Giuseppe T. Cirella * and Barbara Pawłowska

Faculty of Economics, University of Gdansk, 81-824 Sopot, Poland; barbara.pawlowska@ug.edu.pl
* Correspondence: gt.cirella@ug.edu.pl

check for updates

Citation: Cirella, G.T.; Pawłowska, B. Advancements in the Energy Sector and the Socioeconomic Development Nexus. *Energies* **2021**, *14*, 8078. https://doi.org/10.3390/en14238078

Received: 2 November 2021
Accepted: 30 November 2021
Published: 2 December 2021

Publisher's Note: MDPI stays neutral with regard to jurisdictional claims in published maps and institutional affiliations.

Contemporary societies, in conjunction with economies from around the world, show an increasing demand for energy. Energy, the driving force behind development, plays a crucial role in running the modern global system. As a result, the future challenge will be not only to meet the rising demand, but also implement less reliance on depleting fossil fuels, which can degrade the environment. Synergies between fossil fuels and different types of pollution can equate to a widening inequality gap [1], higher economic costs [2], and regulatory oversight into (new) sectors such as "dumping in outer space, renewable energies, environmental information disclosure, and green production technologies" [2]. These constructs can create noticeable differences, from traditional regulatory domains to a new normal where green policy dictates citizenry [3,4]. In consequence, several technologies and interventions have been presented in this Special Issue entitled "Energy Security as a Key Driving Factor for Socioeconomic Development: From Mitigation to Solution." Viable means of reducing and preventing such drivers with significant economic benefits have been documented and predicted. As such, the sustainability of supplied energy requires a reduction in emissions to control the absorption capacity vis à vis the environment. Globally, policymakers have largely recognized the significance of the relationship between energy and economic progress. According to Indriyanto et al. [5], policymakers usually consider the social and economic aspects of energy security in terms of its affordability and accessibility of service. One of the primary concerns of policymakers should be to ensure energy security for its users [6]. The condition of socioeconomic development depends on safe, secure, and sustainable energy at affordable prices. These factors have resulted in an increasing interest to undertake activities that develop renewable energy sources and expand energy alternatives society wide. Energy efficiency is treated as the most cost-effective way to reduce energy demand while maintaining stable economic activity. Some researchers have called this a "fifth fuel", even though it does not have much in common with the traditional sources of energy science [7]. Increasing energy efficiency is an important supportive aspect to solving issues in relation to climate change, energy security, and energy competitiveness. Accordingly, no country can afford to waste energy and must prioritize it if it is to continue to modernize.

Another pressing challenge is rapid economic development in the developing world. This change is highly dependent on energy consumption primarily sourced from fossil fuels [8,9]. This influx mostly considers energy conservation as a perceived additional cost and a lowering of living standards. This standpoint has been considered an approach that denies communities in developing regions the opportunity to improve their living condition and technological progress. Energy poverty represents the situation often observed in developing regions and plays an important part in examining the effect of financial inclusion (i.e., by identifying principal channels) between available funds and available energy [10]. The lack of or limited access to modern energy services, such as electrical power, and the negative effects on well-being associated with energy poverty (e.g., slow economic growth, a low human development index, and high environmental impact) are the effects of such poverty that requisite a mitigation of a solution type of progression.

In line with these goals, this Special Issue collated research from the international scientific community and provided a picture of the global problem of energy security via case studies and technological know-how. Important leading-edge topics include energy efficiency modelling and simulation tools, energy security and development nexus technology, transformative research into renewables, impact of energy poverty on human development and the environment, energy resource studies and exploitation barriers, fossil fuel subsidies and fuel switching, and techno-economic assessment of energy supply solutions and the potential role of carbon-neutral technologies. All these topics have an underlying policy element that require initiatives and recommendations to reduce and limit energy poverty, so as to highlight the energy security socioeconomic nexus of developing solution-based outcomes.

Hoody et al. [11] investigated American-based reduction programs that have overseen utility-sponsored residential energy over the last decade. The programs disclosed important investments in energy efficient appliances and developments. They noted that co-investment by residents of varying socioeconomic backgrounds supported the utility initiatives directed toward behaviour-based energy reduction via "technologies, such as smart meters and smart Wi-Fi thermostats linked to phone apps" [11]. The research sought to explore these programs specifically at low-income residences using peer-to-peer energy education and support. This study correlated previous findings in which qualitative data, obtained from program implementers is a viable starting point for the development of an improved energy design. The design highlighted "grassroots community co-design of the program and community engagement through program implementation to transform energy consumption and behaviours and find energy justice for vulnerable communities" [11].

Janikowska and Kulczycka [12] examined the transitional use of a tool for preventing energy poverty among women in the mining areas of Silesia Region, Poland. The study utilized the Silesia Region as a representative example of an archetypal European mining territory whose economy is primarily based on coal. With job losses on the rise from the mining sector, the study showed demographic and social data of different groups of people, i.e., mainly households inhabited by single women affected by energy poverty. The Just Transition strategy was applied to the situation of women to transition to other (i.e., future) labour markets outside of the mining industry. The process of restructuring the inhabitants was found to be a profound cultural change, which affected "the ethos of conscientious work, reliability, and love of family that is so important to the inhabitants of the region and which is the ethical code of the Silesian population" [12]. Conclusive findings suggested that the process of closing the mining sector should entail "fairness, solidarity, and sustainable development," [12] and be interlinked with assisting such communities with compensation for "incurred costs and losses (including environmental) [as well as to] receive post-industrial infrastructure to be used for scientific, educational, social, cultural, and commercial purposes" [12]. It can be said that, at length, replacing coal should (by default) augment the advancement of renewable energy, energy storage technology, and other operative energy-based technologies that can stabilize the power system.

Rausch and Suchanek [13] identified socioeconomic factors influencing the investment decision in solar power of the prosumer in Germany. They pieced together socioeconomic factors that impact on the investment decision of private households towards investments in small scale solar units throughout the country. With Germany's last nuclear power plant being phased out in 2022 and its coal-fired power plants being turned off in 2038, legislators have mandated renewable energy alternatives to close the gap that fossil fuels and nuclear power will leave behind. As such, a portion or share of the prescribed "renewable energies could be [sourced] from private households that mainly invest in small scale solar" [13]. This study examined Germany's energy transition to stimulate investment decisions of private households. Secondary socioeconomic data from 2005 to 2018 found, via a factor analysis of identified latent variables, that five factors have an impact on the investment decisions of prosumers: socioeconomics, urbanization, education, scale of industrialization, and the birth-to-death rate variance. They concluded that investments from prosumers are mostly found in the southern parts of the country "where the quality of life is high and many inhabitants live [on] their own properties in rural areas" [13]. In contrast, inhabitants living in urban areas with a "low level of freestanding houses and a lower level of property ownership" [13] did not invest in (alternative) home-based solar—instead opting for traditional energy means. A pivotal, transitory phase for legislators will be when municipal authorities, in conjunction with the participation of the inhabitants, deindustrialize the country's energy from coal-fired power plants in a post-2038 Germany.

Czermański et al. [14] proclaimed that container shipping is the largest producer of emissions within the maritime shipping industry. They formulated an energy consumption approach to estimate air emission reductions in container shipping as a means of measuring ship emission levels. This research is linked to the International Maritime Organization's MARPOL Annex VI application of Tier III requirements, the Energy Efficiency Design Index for new ships, and the Ship Energy Efficiency Management Plan for all ships. It amalgamated findings that can assist policy formulation germane to energy consumption by estimating the volume of sulfur oxide, nitrous oxide, particulate matter, and carbon dioxide emitted from container ships (i.e., via logged data and average vessel speed records generated by the automatic identification system). The research was mapped using geographic information systems and framed empirical findings "to estimate ongoing emission reductions on a continuous basis […] to fill data gaps where needed, as the latest worldwide container shipping emissions records date back to 2015" [14]. The study reinforces early stage detection of environmental impacts and helps to adopt the greatest potential for emission reductions in terms of location.

Fu et al. [15] examined the effects of regional innovation capability on green technology efficiency of China's manufacturing industry between 2011 and 2017 in A-share listed enterprises. This study highlighted the innovation capabilities of the local manufacturing industry to achieve green technology and sustainable development initiatives. The research took an explicit look at whether "regional innovation capabilities can promote the improvement of green technology manufacturing efficiency [and found] a significant spatial correlation between [the two as] prevalent within spatial heterogeneous bounds" [15]. In a geographical context, it was illustrated that regional innovation capability was strongest in eastern China, in which human capital and government revenue aided in advancing the green technology sector. Green technology can be seen as an alternative practice of facilitating cleaner energy and part of the transitory solution to energy security.

Cirella et al. [16] presented an expository essay that looked at the rural-to-urban transition and correlated it with urban energy demands. The essay examined three distinct themes to developing awareness for urbanization: internal urban design and innovation, technical transition, and geopolitical change. Over the last 30 years, the authors argued that "the urban population boom continues to pressure the energy dimension with heavily weighted impacts on less developed regions; [moreover, unsustainable] urban energy will need to reduce resource inputs and environmental impacts" [16]. It was noted that a decoupling of economic growth from energy consumption will also need to be facilitated regardless of fossil fuel usage (i.e., the preferred method of energy for cities). They stated an "increased understanding is emerging that sustainable energy forms can be implemented as alternatives" [16]. The key to this future transition will be the will to invest in renewables (i.e., solar, wind, hydro, tidal, geothermal, and biomass), efficient infrastructure, and smart eco-city designs. The essay clarified how the technical transition of energy-friendly technology can be implemented into the overall energy mix and how smart electricity-based storage grids—with artificial intelligence—can aid at the international level as well as enforce an energy re-shift to a better human-energy-oriented relationship.

The contributions to this Special Issue presented a broad view to the advancement of the energy sector in correlation to the socioeconomic development nexus. Focusing solely on savings or increasing the efficiency of energy use will not be sufficient [17,18]. In the near future, it will be necessary to work towards a planet-friendly energy mix approach. Improvements to the production process, modernization of equipment and buildings, and the introduction of new technologies will all be vital activities aimed at improving the energy efficiency of the economy. This idea is centred on providing the same number of products or services using less energy and less raw materials to produce it. The effect should reduce polluting emissions and increase the energy security of the state. The fundamentals of energy efficiency policy spurs from this design (i.e., concept)—to the betterment of assessing and improving energy efficacy—so as not to impede on economic growth or economic competitiveness. Energy efficiency targets may be achieved by applying market measures that ensure economic benefits by optimizing technological-economic processes, while considering the complexity of the energy efficiency issue inclusive of the environment. This, however, cannot be attained without combining the efforts of all sectors of the economy and individual users. As such, changing final user behaviour is an essential part of the energy transition process. Synergy, obtained in this way, should allow for the achievement of ambitious energy goals, the reduction of environmental degradation, and the assurance of energy security—sector-wide.

Author Contributions: Conceptualization, investigation, resources, writing—original draft preparation, writing—review and editing, G.T.C. and B.P. All authors have read and agreed to the published version of the manuscript.

Funding: This research received no external funding.

Informed Consent Statement: Not applicable.

Data Availability Statement: Not applicable.

Acknowledgments: We are very grateful to all the authors for submitting their work to this Special Issue. We would also like to acknowledge the Energies editorial team and reviewers for enhancing the academic quality, rigor, and impact of all the contributions.

Conflicts of Interest: The authors declare no conflict of interest.

References

1. Perera, F. Pollution from Fossil-Fuel Combustion is the Leading Environmental Threat to Global Pediatric Health and Equity: Solutions Exist. *Int. J. Environ. Res. Public Health* **2018**, *15*, 16. [CrossRef]
2. Hao, F.; Van Brown, B.L. An Analysis of Environmental and Economic Impacts of Fossil Fuel Production in the U.S. from 2001 to 2015. *Soc. Nat. Resour.* **2019**, *32*, 693–708. [CrossRef]
3. Barbier, E. How is the Global Green New Deal going? *Nature* **2010**, *464*, 832–833. [CrossRef]
4. Bouzarovski, S.; Thomson, H.; Cornelis, M. Confronting Energy Poverty in Europe: A Research and Policy Agenda. *Energies* **2021**, *14*, 858. [CrossRef]
5. Indriyanto, A.R.S.; Fauzi, D.A.; Firdaus, A. The Sustainable Development Dimension of Energy Security. In *The Routledge Handbook of Energy Security*; Sovacool, B.K., Ed.; Routledge: Abingdon, UK, 2010; pp. 96–112; ISBN 978-0-203-83460-2.
6. Umbach, F. Global energy security and the implications for the EU. *Energy Policy* **2010**, *38*, 1229–1240. [CrossRef]
7. Yergin, D. *The Quest: Energy, Security, and the Remaking of the Modern World*; Penguin Books: New York, NY, USA, 2013.
8. Awodumi, O.B.; Adeleke, A.M. Non-Renewable Energy and Macroeconomic Efficiency of Seven Major Oil Producing Economies in Africa. *Zagreb Int. Rev. Econ. Bus.* **2016**, *19*, 59–74. [CrossRef]
9. Awodumi, O.B.; Adewuyi, A.O. The role of non-renewable energy consumption in economic growth and carbon emission: Evidence from oil producing economies in Africa. *Energy Strateg. Rev.* **2020**, *27*, 100434. [CrossRef]
10. Koomson, I.; Danquah, M. Financial inclusion and energy poverty: Empirical evidence from Ghana. *Energy Econ.* **2021**, *94*, 105085. [CrossRef]
11. Hoody, J.; Robertson, A.G.; Richard, S.; Frankowski, C.; Hallinan, K.; Owens, C.; Pohl, B. A Review of Behavioral Energy Reduction Programs and Implementation of a Pilot Peer-to-Peer Led Behavioral Energy Reduction Program for a Low-Income Neighborhood. *Energies* **2021**, *14*, 4635. [CrossRef]
12. Janikowska, O.; Kulczycka, J. Just Transition as a Tool for Preventing Energy Poverty among Women in Mining Areas—A Case Study of the Silesia Region, Poland. *Energies* **2021**, *14*, 3372. [CrossRef]
13. Rausch, P.; Suchanek, M. Socioeconomic Factors Influencing the Prosumer's Investment Decision on Solar Power. *Energies* **2021**, *14*, 7154. [CrossRef]
14. Czermański, E.; Cirella, G.T.; Oniszczuk-Jastrząbek, A.; Pawłowska, B.; Notteboom, T. An Energy Consumption Approach to Estimate Air Emission Reductions in Container Shipping. *Energies* **2021**, *14*, 278. [CrossRef]
15. Fu, Y.; Supriyadi, A.; Wang, T.; Wang, L.; Cirella, G.T. Effects of Regional Innovation Capability on the Green Technology Efficiency of China's Manufacturing Industry: Evidence from Listed Companies. *Energies* **2020**, *13*, 5467. [CrossRef]

16. Cirella, G.T.; Russo, A.; Benassi, F.; Czermański, E.; Goncharuk, A.G.; Oniszczuk-Jastrzabek, A. Energy Re-Shift for an Urbanizing World. *Energies* **2021**, *14*, 5516. [CrossRef]
17. Welsby, D.; Price, J.; Pye, S.; Ekins, P. Unextractable fossil fuels in a 1.5 °C world. *Nature* **2021**, *597*, 230–234. [CrossRef] [PubMed]
18. Kirsch, S. Running out? Rethinking resource depletion. *Extr. Ind. Soc.* **2020**, *7*, 838. [CrossRef]

 *energies*

Article

A Review of Behavioral Energy Reduction Programs and Implementation of a Pilot Peer-to-Peer Led Behavioral Energy Reduction Program for a Low-Income Neighborhood

Jennifer Hoody [1,*], Anya Galli Robertson [2], Sarah Richard [3], Claire Frankowski [1,3], Kevin Hallinan [1,3,*], Ciara Owens [3] and Bob Pohl [3]

[1] Department of Mechanical and Aerospace Engineering, University of Dayton, Dayton, OH 45469, USA; cfrankowski3@gmail.com
[2] Department of Sociology, Anthropology, and Social Work, University of Dayton, Dayton, OH 45469, USA; agallirobertson1@udayton.edu
[3] Clean Energy 4 All, Dayton, OH 45410, USA; Sarah.n.richard19@gmail.com (S.R.); ciara.joyvin@gmail.com (C.O.); pohlbob@gmail.com (B.P.)
* Correspondence: jenn.hoody@gmail.com (J.H.); kevin.hallinan@udayton.edu (K.H.)

Citation: Hoody, J.; Galli Robertson, A.; Richard, S.; Frankowski, C.; Hallinan, K.; Owens, C.; Pohl, B. A Review of Behavioral Energy Reduction Programs and Implementation of a Pilot Peer-to-Peer Led Behavioral Energy Reduction Program for a Low-Income Neighborhood. *Energies* **2021**, *14*, 4635. https://doi.org/10.3390/en14154635

Academic Editors: Barbara Pawlowska and Giuseppe T. Cirella

Received: 30 May 2021
Accepted: 26 July 2021
Published: 30 July 2021

Publisher's Note: MDPI stays neutral with regard to jurisdictional claims in published maps and institutional affiliations.

Abstract: Utility-sponsored residential energy reduction programs have seen rapid advancement in the Unites States (US) over the past decade. These programs have particularly emphasized investments in energy efficient appliances and enveloped improvements. They have generally required co-investment by residents and, as a result, have mostly reached medium to high-income residents, with low income residences, in effect, supporting the utility investments through higher energy costs. Additionally, utility initiatives directed toward behavior-based energy reduction have reached residences with more advanced technologies, such as smart meters and smart Wi-Fi thermostats linked to phone apps, technologies generally not present in low-income residences. This research seeks to inform development of behavior-based energy reduction programs aimed specifically at low-income residences, premised on peer-to-peer energy education and support. It focuses on the design and implementation of a pilot program for 84 low-income residences in a medium-sized Midwestern US urban neighborhood, followed by measurement of realized energy savings and assessment of program outcomes through surveys of resident participants and interviews with program implementers. Only 21 residences provided an initial response to outreach. Of these, only 11 participated, and of these, energy savings were, in general, modest. However, evidence based upon other research and qualitative data obtained from program implementers and participants is presented in this study for the development of an improved design. The improved design emphasizes grassroots community co-design of the program and community engagement through program implementation to transform energy consumption and behaviors and find energy justice for vulnerable communities.

Keywords: energy burden; peer-to-peer; energy behavior; energy justice; low-income; underserved communities; energy savings

1. Introduction

Research from the scientific community attests that climate change is a paramount concern in the contemporary world and has identified humankind as a primary catalyst of the rising risks. According to the Intergovernmental Panel on Climate Change's 2018 Special Report, global temperatures have risen approximately 1 °C from human activity since preindustrial times [1]. Modern society remains dependent on the production and consumption of substantial amounts of energy, and the energy sector plays a vital role in the everyday life of a large share of the global population as well as in the economy. The residential sector is responsible for a significant portion of energy consumed and the resulting greenhouse gases emitted; nearly 20% of the 5130 million metric tons of carbon dioxide emitted by the US in 2019 was from the residential sector [2]. The deleterious impacts from emissions are spatially uncontrollable, but vulnerable populations have been the most acutely affected [3,4]. Thus, the production, distribution, and consumption of energy is a global concern for environmental and social justice.

Energy justice looks at the energy sector from a social justice perspective to expand the scope of energy beyond economic and societal benefits. It analyzes and reveals the human costs of energy and the injustices that have resulted from the increased demand and reliance on energy [5]. The principle of energy justice has numerous definitions, but generally, the principle stems from the theories of distributive, procedural, and recognition justice [6]. According to Sovacool and Dworkin (2015), energy justice is defined as the following: *"The right of all to access energy services, regardless of whether they are citizens of more or less developed economies. It encompasses how negative environmental and social impacts related to energy are distributed across space and time, including human rights abuses and the access that disenfranchised communities have or should have to remedies [5]"*.

There are two prominent ways in which distributive justice applies to energy justice: (1) the spatial and temporal location of energy infrastructure and access to energy, and (2) the benefits and costs that accompany the production, distribution, and consumption of energy. The procedural component of energy justice concerns energy policy and decision-making processes that are just and transparent such that individuals have equal input and are equally represented and considered. Lastly, energy justice includes recognition justice, which is the theory emphasizing the necessity of properly identifying all forms of injustice within the energy sector and, therefore, is essential for achieving procedural justice [5].

Energy insecurity and energy burdens are two types of injustices that exist in the energy sector. Energy insecurity refers to energy as an unstable and unreliable resource for vulnerable populations that are physically or financially disadvantaged [4,7,8]. According to the US Energy Information Administration's (EIA) Residential Energy Consumption 2015 Survey, out of a total of 118.2 million US households, 37 million were energy insecure, with 25.3 million reducing food purchases or forgoing the purchase of medicine to pay utility bills, 12.8 million living in unhealthy temperature living environments, and 17.2 million being disconnected from energy access [9].

In contrast to energy insecurity, energy burden pertains exclusively to financial in-equality in energy, expanding beyond financial and physical inequalities and disadvantages. Specifically, energy burden is the percentage of a household's gross annual income spent on utility bills. This is a social injustice because underprivileged populations endure dispro-portionately high energy burdens when compared with their counterparts [10,11]. In the US, 25.8 million low income households experience an average energy burden of approxi-mately 8.1%, over 3.5 times greater than that of non-low-income households, whose energy burden is approximately 2.3% [11]. Furthermore, energy burdens are disproportionately higher for minorities and other marginalized populations. According to the American Council for an Energy-Efficient Economy's 2021 assessment of household energy burdens by Drehobl et al. in comparison with 1% of non-low-income and 9% of non-Hispanic white households, 21% of black households experience severe energy burdens, which is defined as energy burdens where households spend at least three times more of their income on utility bills than median households [11]. The energy burdens endured by households are further exacerbated by unequal access to clean energy offerings [12].

Because energy insecurity and high energy burdens most severely impact financially disadvantaged populations and communities of color, the energy injustice these households face is oftentimes coupled with food and housing insecurity. In 2015, the EIA estimated that, of the aforementioned 25.3 million households that forwent food and medicine to pay utility bills, 7 million were faced with this decision on a monthly basis [9]. The aggregate of these injustices not only amplifies hardships but also contributes to and perpetuates intergenerational injustices in historically segregated and disenfranchised neighborhoods [4]. This can force households to live in uncomfortable, unsafe, stressful, and unhealthy conditions in order to pay their utility bills, ultimately increasing health risks and amplifying the burdens households endure [4,8].

Energy behavior, defined as the habits, motivations, and values associated with energy consumption, is an important component of energy efficiency. However, it is often neglected and not a central focus of energy reduction initiatives. Incorporating energy behavior education and tactics into energy cost reduction programs can open the door to significant energy savings [13]. This will be exceptionally beneficial for low-income populations who are not able to make energy efficiency upgrades and repairs and who are not able to receive sufficient assistance from pre-existing programs. In a study conducted by Ouyang and Hokao (2009), residential occupants were provided education on a set of measures for the use of electric appliances to improve energy saving behaviors (e.g., refrigerating foods once they have cooled completely), and the study concluded that the adoption of energy saving behaviors has the potential to reduce energy consumption by nearly 14% in the residential sector [13]. The potential for savings among low-income populations, however, has not been well-documented. Nevertheless, it is certain that improved energy behavior has the potential to advance efforts toward an energy-just world through methods beyond energy efficiency and energy assistance programs alone.

2. Background

2.1. Energy Poverty

Energy poverty, commonly referred to as energy insecurity and fuel poverty, is a multidimensional facet of energy injustice that has gained increased attention by scholars and researchers [7,14]. Broadly defined as the deprivation of energy, energy poverty has numerous meanings based on the application, in which the reality of energy deprivation differs between and within developed and developing economies [8]. In 2017, nearly one billion people lacked access to energy globally [15]. Of this, an estimated 87% of the world population without electricity are in rural areas [16] and 200 million people in developed economies suffer from energy poverty [17]. Energy poverty is, thus, a complex concern that requires a comprehensive understanding of energy deprivation from a global, national, and local perspective.

Acknowledgement and action against energy poverty first began as the concept of fuel poverty in the United Kingdom (UK) in response to rising energy costs in the 1970s. The early theory of fuel poverty emphasized household incomes, high energy costs, and poor domestic energy efficiency as causes of energy deprivation and formally defined fuel poverty as households spending more than 10% of their income on energy [8,10,17]. Since then, the understanding of energy poverty has evolved and is also typically associated with lack of access to modern energy and unreliable energy services [10,17]. These understandings alone, however, fail to acknowledge vital components of the reality of energy poverty. Energy poverty cannot be misidentified and merely associated with general income poverty because it encapsulates more than being able to afford energy and includes external factors such as physical personal and infrastructural limitations, medical conditions, and social, economic, and political factors [4]. Furthermore, financial development, policy, infrastructure, food and housing insecurity, physical and mental health, institutional racism, etc. are critical characteristics that must also be considered when analyzing energy poverty [7,8,18]. Acknowledging this nexus of energy poverty and the aforementioned systems and injustices is essential as focus turns toward combating energy poverty.

In the US, there are two federally funded energy assistance programs that are intended to address these inequalities: The Low Income Home Energy Assistance Program (LIHEAP) and The Weatherization Assistance Program (WAP). LIHEAP is offered by the Office of Community Services through the Department of Health and Human Services, which serves to help low-income households meet their energy needs. Services offered through these programs include bill payment and energy crisis assistance, energy-related home repairs, and weatherization, which is the process of making improvements and upgrades to increase a home's energy efficiency and resistance to weather changes [19]. WAP is operated through the Department of Energy, offering eligible households weatherization improvements and upgrades to increase energy efficiency and, thus, reduce energy costs [20]. These programs aim to alleviate energy burdens, reduce utility bills for energy insecure homes, and improve the health and safety within a home.

There exists, however, a significant gap between the capacity of these programs and the need for assistance. According to Bednar and Reames (2020), LIHEAP provides assistance to approximately only 25% of eligible households per year, and out of nearly 40 million eligible households, WAP has only been able to weatherize 7 million households [10]. Barriers that prevent greater access to these programs include available funding, state level priorities, and the high demand for weatherization, which is much greater than the rate at which weatherization improvements can be implemented [4,11]. With the programs' focus primarily on momentary relief, they also do not provide sustainable solutions to eliminate energy poverty. In addition, if residents fail to pay their utility bills, they are required to pay back all energy assistance benefits they have received, thus, when eligible, low-income residents are often deterred from accepting energy assistance through LIHEAP. While LIHEAP was intended to be an emergency option for low-income residents, the reality is that residents enrolled in this program remain reliant on it in perpetuity. This assistance does not address the systemic conditions that result in high energy burdens and, with performance measures failing to measure long-term impacts, has not been shown to reduce the incidence of the non-payment of energy bills [4,10,21]. In contrast to energy bill assistance, weatherization and improved household energy efficiency has shown greater value in addressing energy poverty through energy cost savings, positive health and safety impacts, employment opportunities, and a foundation for continued energy efficiency upgrades [10] and has the potential of reducing energy burdens by 25% for low-income households [11]. Weatherization through WAP, however, is significantly underfunded in comparison to the allocated funding for energy bill assistance through LIHEAP ($3 billion USD compared to $0.4 billion USD, respectively) [10], and at the current rate, weatherizing all eligible households through WAP will take an estimated 360 years [11].

It is evident that greater action is needed to address energy poverty both within the US and across the globe. Despite the intensifying concerns and impacts of climate change that has led to a greater global push for sustainable development, there has been a lack of recognition and response to energy poverty [10]. For instance, the United Nations' Sustainable Development Goal 7 sets to achieve universal access to affordable, reliable, sustainable, and modern energy by 2030; yet, in a 2018 report of 46 national reviews, there was little acknowledgement of energy poverty [16]. To shift this trend and increase response to energy poverty, it will be essential that action taken includes measures beyond income poverty and includes factors such as financial development, advancement of renewable energy systems, and the health impacts of the lack of energy access [10,14,15]. In a study conducted by Nguyen et. al., 2021, the relation between financial development and energy poverty was analyzed for 65 economies across the globe. The study concluded that to address energy poverty, low and lower-middle income economies should focus efforts on the development of the financial sector for market-based support and on government policy that supports renewable energy, whereas upper-middle income economies should emphasize policy and regulations on the financial sector that push for greater sustainable development [14]. Global action must be methodologically established to acknowledge the nuanced reality of energy poverty, such as the dichotomy between developing and developed economies.

To fight energy poverty, it is critical to expand the acknowledgement and understanding of its complex and multifaceted nature. Inadequate and lack of action stems from the non-recognition and disrespect of energy poverty, consequently suppressing it in political debate and action [7]. Steps must therefore be taken to confront stereotypes that generalize energy poverty based on certain characteristics and demographics, such as income, and that recognize energy access as a human right that is influenced by the design and implementation of policy and infrastructure [7,8]. While numerous factors of energy poverty are beyond the control of individual households, the present study takes a localized and household level approach in order to refute generalized notions about those experiencing financial and energy poverty. This study will contribute to existing literature by gaining perspectives and insight on energy poverty from a household level. Additionally, it will address energy poverty by increasing access to energy education and equipping households with information and resources to better understand and take control of their energy behaviors.

2.2. Values and Motivation of Energy Consumption

A fundamental step in understanding the extent to which individuals adopt environmentally conscious and energy saving behaviors is to first analyze values and motivations. Previous studies on residential energy consumption reveal an exhaustive list of factors that influence values and motivations associated with energy savings behaviors. Hence, the findings discussed in this study do not provide a definitive approach to understanding values and motivations but instead serve as a guide and basis of factors to be considered.

The intentions behind individual energy behaviors take numerous forms. Lindenberg and Steg (2007) propose that behaviors and actions are driven by goals and how such goals are framed [22]. This theory postulates three goal frames: gain, normative, and hedonic. Gain goal frames are driven by protection and advancement, normative goal frames are driven by what is proper and acceptable, and hedonic goal frames are driven by the desire to feel better at a given moment. When applying these to environmentally conscious behavior, it is suggested that hedonic goal frames impact behaviors the strongest.

Intentions and goals alone, however, do not provide enough context for understanding environmentally and energy conscious behavior. To analyze the gap between intent and action, additional factors such as education, skills, and demographics must also be considered. Hines et al., 1987, indicated that in order for positive intention to lead to environmentally conscious behavior, cognitive knowledge and skills are essential [23]. The most successful results were seen when individuals were not only aware of the problem and actions they could take, but when they were equipped with the skills to effectively and successfully act. Furthermore, when the desire and intent to act in an environmentally conscious manner was lacking, the ability to act was more likely impacted by situational factors such as economic and social constraints. According to a study completed by Poortinga et al., 2004, *"attitudinal variables explained a mere 2% of variation in home energy use, the variation explained increased to 15% after taking into account several socio-demographic variables"* [24]. This research, therefore, conveys the interconnected relationship between personal intention and desire, accessibility to knowledge and skills, and socio-demographics and the complexity of understanding and achieving environmentally conscious behaviors.

The sense of personal and social influence over environmentally conscious behaviors is a factor that must also be evaluated. In a meta-analysis completed by Hines et al., 1987, it was revealed that self-blame and internal locus of control tend to lead to and be associated with environmentally conscious behaviors [23]. By taking personal responsibility, individuals are able to see and acknowledge that their actions are effective and impactful. In addition to internal influence, when individuals are exposed to social norms that promote such environmentally conscious behavior, their likelihood to engage in such behavior increases further, and they are more apt to modify current behaviors.

A behavioral nudge is another method of cognitive behavioral change that has implications for energy behaviors. According to Thaler and Sunstein (2008), a nudge is *"any aspect of the choice architecture that alters people's behavior in a predictable way without forbidding any options or significantly changing their economic incentives"* [25] (p. 6). There are numerous forms of nudges in the context of behavioral change and, as described by Thaler and Sunstein, can take the form of social and emotional nudges when analyzed from an environmentally conscious perspective. In a study of 300 households in San Marcos, California, households were provided energy consumption data for their individual household and the average for their neighborhood average [25]. For half of the households, this information also included icons indicating whether their energy usage was socially acceptable or not (i.e., icons portraying happy or unhappy emotions, respectively). These households were provided a social and emotional nudge.

After receiving information merely on previous energy consumption, above-average energy consuming households decreased their energy usage and below-average energy consuming households increased their energy usage. However, the outcome was more advantageous for the households who also received an emotion-icon: above-average consuming households substantially decreased their energy usage and below-average consuming households did not increase their energy usage, in contrast to households who did not receive an emotion-icon. By utilizing feedback to make energy consumption tangible and connecting to emotions and societal values, this study shows the value in leveraging social and emotional nudges to promote energy conscious behavior [25]. However, limited research and literature exists on applying behavioral nudges in an environmental and energy saving context [26].

These research findings indicate the complex and dynamic nature of energy behavior that is dependent upon a multitude of factors. The aforementioned studies primarily focused on a range of demographics and are not indicative of how values and motivations may differ for low-income communities specifically. For instance, low socioeconomic status may have different needs or priorities and, in many cases, may be unable to act upon environmentally conscious and energy savings values. Nonetheless, these insights provide an understanding of how values and motivations are influenced by internal and external factors and how they shape energy behavior.

2.3. Peer-to-Peer Education

Peer-to-peer education is a method in which a representative, educator, mentor, or coach of a specified program is of the same or similar background as the participant [27]. This method has been implemented across a multitude of fields and demographics, but little to no research exists applying this method in underserved communities to modify energy behavior and decrease energy consumption.

The understanding behind the value of peer-to-peer methodology can be explained from a psychological standpoint. In a study analyzing the impact of peer teaching in medical education, psychologists suggest that the success of such teaching is linked to two factors: cognitive and social congruence [28]. From a cognitive perspective, learning takes place when new information is introduced to the brain and relationships and networks are established with pre-existing knowledge to adopt the new information. Cognitive congruence implies that an individual is more apt to introduce information to their peer by minimizing the gap between new and pre-existing knowledge. In addition, social congruence explains that peer-to-peer education is effective because peers are more vulnerable and less anxious with someone they relate to as compared to figures of authority and perceived superiority, ultimately increasing confidence and the ability to learn.

The efficacy of peer-to-peer education has been studied in fields such as health, nutrition, and education to analyze and validate the benefits of peer-to-peer indicated by these psychological explanations. In one study completed at the University of California, San Francisco, the impacts of peer education and coaching among low-income patients with diabetes were investigated [29,30]. Patients were recommended by clinicians to partake in training to become peer health coaches for patients with similar health backgrounds to determine if the role of a peer health coach would aid in the reduction of hemoglobin A1C (HbA1C) levels. While the retention of the peer health coaches decreased by over half from enrollment to the completion of the study, data from the training sessions revealed that 86.5% completed the training and 81.3% passed the final written and oral exams administered prior to health coaching. Among the patients who went through training, 28.1% had graduated from college and 25% had not completed high school [29]. Despite these factors, after six months of peer coaching, there was a significant reduction in HbA1C levels among patients receiving peer education support when compared with patients who did not participate [30]. This study revealed that lower socioeconomic status individuals can successfully acquire necessary knowledge and skills and serve as effective peer educators.

Furthermore, peer-to-peer research has been conducted in nutrition education in low-income communities. Developed by California's Public Health Department Nutrition Program, two programs, Head Start and Parents as Teachers, were created to increase the knowledge and improve behaviors and intentions for healthy and low-cost nutrition among low-income parents [31]. The programs consisted of two nutrition classes offered to parents that were taught by fellow parents. To measure the effectiveness of the program and of the peer education method implemented, questionnaires were administered prior to and after the completion of the classes to gather data regarding content of the class, as well as demographics. The results revealed that not only were parents overwhelmingly satisfied with the courses, but it also showed an increase in knowledge. According to the pre-class questionnaires, only 40.2% of participants were able to correctly identify low-fat foods, which increased to 95.1% correct identification post-class [31]. This program also revealed that optimal results were achieved when the peer-parent-teachers contributed to the structure of the program, which indicated an increase in commitment and personal investment. While the program did not study the long-term impact, it, nonetheless, confirmed that peer-to-peer education among low-income parents can successfully increase knowledge and intentions centered around healthy eating.

The use of peer-to-peer education and support has also played a prominent role among a multitude of services for low-income pregnant mothers. People's Equal Action and Community Effort Incorporated (PEACE) and Early Head Start (EHS) are federally funded services that serve pregnant women and families with young children in Onondaga County, New York, which, at the time of the study, had one of the highest infant mortality rates in the country [32]. In addition to home visits the program already provided, the Pregnancy Care Campaign (PCC) was created. This program revolved around a variety of events where participating expecting mothers were educated and motivated to live healthier pregnancies through interactions with professional educators and peer mothers. A primary goal of the PCC events was to allow the participating mothers to open up with other mothers in similar situations based upon the idea that *"the knowledge of another person's experience helps inform one's own decision especially in making personal choices"* [32]. One study of the campaign followed first-year participating mothers and found that there were no low-weight births or premature infants and that there was an increase in prenatal care among the mothers. Thus, this provides further confirmation on the role peer-to-peer education and mentoring can have among low-income communities and individuals.

The analysis of peer-to-peer based diabetes, nutrition, and pregnancy programs validates that peer-based behavior education can render change among low-income communities and individuals. The research posed here investigates if the same methodology can be used to realize significant energy cost savings through behavioral modifications, as there appears to be limited to no prior research investigating this application. Specifically, the present study outlines the development of a peer-to-peer energy reduction program for underserved communities, the preliminary results from a pilot program, and the knowledge gained during the pilot program for an improved program design, with the aim that these findings will amplify the impact of this program framework for future applications.

3. Case

3.1. Case Study Overview: Pilot Peer-to-Peer Energy Reduction Program

This project focuses on a pilot program conducted by a clean energy non-profit organization whose goal is to achieve energy and cost savings for low-income communities, specifically, in the Twin Towers neighborhood in East Dayton. At the time of the pilot program, Twin Towers was composed predominantly of members of white, black, Asian, Hispanic or Latin, and American Indian communities [33]. Many households within the neighborhood live in financial poverty with over 50% of all households and nearly 67% of female led households living in government defined poverty and approximately 65% of the families renting their home [34]. Between 2009 and 2013, 84 rent-to-purchase homes were built in Twin Towers to provide affordable housing as part of the Low-Income Tax Credit program, a tax credit for affordable housing directed toward low-income individuals in the US (Theoretically, the residents are eligible to purchase their home after a 15-year time period in which tax benefits can be obtained by equity investors. Having lived there for a long time, the residents would have accrued equity in the house, making purchase more feasible. However, a majority of the annual earnings of those living in the homes is less than 2/3 of the median income and much of this housing is generally transient with few residents living in the houses for more than five years. Thus, homeownership is rarely attained).

The housing manager of the 84 homes required all residents to sign a utility release form for utility information to be obtained to understand the financial situation of each household. Through these releases and a partnership with the providing energy utility, monthly energy consumption data for each participating household were made available for this study. These 84 homes are similarly constructed three and four bedroom models with a floor area of 110 and 140 m^2, respectively, built for affordability with relatively high energy efficiency characteristics [35]. The electrical energy consumption of the 84 residences should have been less than the national average (approximately 10,000 kWh/year [9]), given that the floor area of the residences was about a third less than the average US residence as well as the fact that the houses were insulated better than average and included only low energy lighting. However, this was not the case. The mean annual electrical energy consumption of all residences was 10,300 kWh, slightly above the national average. Moreover, there was wide variation in energy consumption. Average annual energy consumption ranged from 3600 kWh/year to as high as 24,000 kWh/year (standard deviation from the mean was 2614 kWh/year). Given the consistency of the housing set, the wide variation in energy consumption is only explainable from energy use behavior differences.

The original goal for the clean energy non-profit was to make an initial investment and install Wi-Fi, smart Wi-Fi thermostats, and solar panels at no cost to the residents in the 84 homes. Through these investments, the intent was to reduce energy costs by an estimated 10% and 50% in the short-term and long-term, respectively. To achieve this, the non-profit would utilize smart Wi-Fi thermostat data, building energy and geometrical characteristics data, occupancy data, and energy and water consumption data to generate machine learning models predicting the monthly energy consumption. These models would provide continuous data needed to analyze energy efficiency and identify areas for improvement, along with estimates of the financial value of the investments. In addition to these measurable goals, the program would offer employment opportunities for community members through the role of a peer-to-peer (P2P) energy educator, the main source of communication with the community and participating residents. A final long-term goal was to hire and train community members to complete energy efficiency upgrades and installations. Thus, the program was designed to lighten the burden of high utility bills and provide employment opportunities for the respective community. A summary of characteristics and demographics of the pilot program can be found in Table 1.

3.2. Pilot Program

The present study investigates the role energy behavior has in promoting energy savings among low-income residents through a unique approach that utilizes peer-to-peer education. Through research and analysis of previous studies, an action plan for peer-to-peer education was formulated, detailing outreach to the community to invite residents to participate, hiring and training a P2P energy educator, managing Wi-Fi and smart Wi-Fi thermostat installations, delivering energy education, and distributing feedback to participants.

To educate and enact energy saving behaviors, a P2P energy educator was hired and trained to work with participants in the pilot program. While this role is intended to be filled by an individual from within the community, the first P2P energy educator was not a resident in the Twin Towers neighborhood. Nonetheless, they had shared lived-experiences and a deep understanding of the lifestyle of those they would be working with. They also had valuable experience in community development, which was a driving factor as to why they were chosen to fill this position for the pilot program. The goal was that they would use their experiences to connect with participants and establish a firm foundation for the position to be assumed by community members in the future.

Table 1. Summary of Program Details.

Program Details	Description
Neighborhood Location	Twin Towers neighborhood in Dayton, Ohio
Financial support	Ohio Housing Finance Authority
Residence type	Low-income rent-to-purchase housing
Neighborhood median income	$32,542 [33]
Average household size	7.8 People [33]
Neighborhood racial demographics	White, Black, Asian, Hispanic or Latino, two or more races, American Indian [33]
Participants racial demographics	Black, White
P2P energy educator racial demographics	White
Program developers' racial demographics	White

Responsibilities of the P2P energy educator included contacting residents interested in participating, installing the noted thermostats in homes, communicating and forging relationships with participants, and educating and collaborating with participants to achieve energy savings. In addition to the P2P energy educator, there was a technical undergraduate intern. This individual worked alongside the P2P energy educator to facilitate the installation of the smart Wi-Fi thermostats, assist with the energy education process, and be a technical resource for the households and P2P energy educator.

Once the program structure was developed and the P2P energy educator and technical intern positions were filled, the first step of implementation was to inform residents in the Twin Towers neighborhood about the program (Figure 1). The 84 rent-to-purchase homes were the focus of the pilot program because, as previously discussed, the homes were built with similar structural and energy efficient characteristics, yet there were significant discrepancies in annual energy consumption. Thus, there was opportunity for behavior-based energy savings among these houses. To contact residents, program flyers were mailed to each resident with program details and a form to register. Additionally, the P2P energy educator and technical intern expanded their outreach by going door to door to familiarize residents with the program. Out of the 84 households, 21 initially signed up for the program, and, ultimately, 11 responded to follow-up communication and participated in the pilot program.

The P2P energy educator then followed-up with participating residents to introduce themselves and begin the process of installing Wi-Fi and smart Wi-Fi thermostats in each home. The Wi-Fi was to be installed and supported at no cost to residents. Several households, however, already had Wi-Fi, thus these households instead received a monthly gift card to a local grocery store to provide equal value to them. Before the P2P energy educator and technical intern began the energy education process, there was a period of approximately one month to collect baseline data for the purpose of comparing energy consumption before and after energy education. The baseline data included the historical energy consumption for the twelve months prior to the study. While baseline data was being collected through a partnership with the energy utility providing service to the residences, the P2P energy educator maintained regular communication with participants to further establish relationships and trust and to check-in and trouble-shoot issues they experienced with their newly installed Wi-Fi and thermostats.

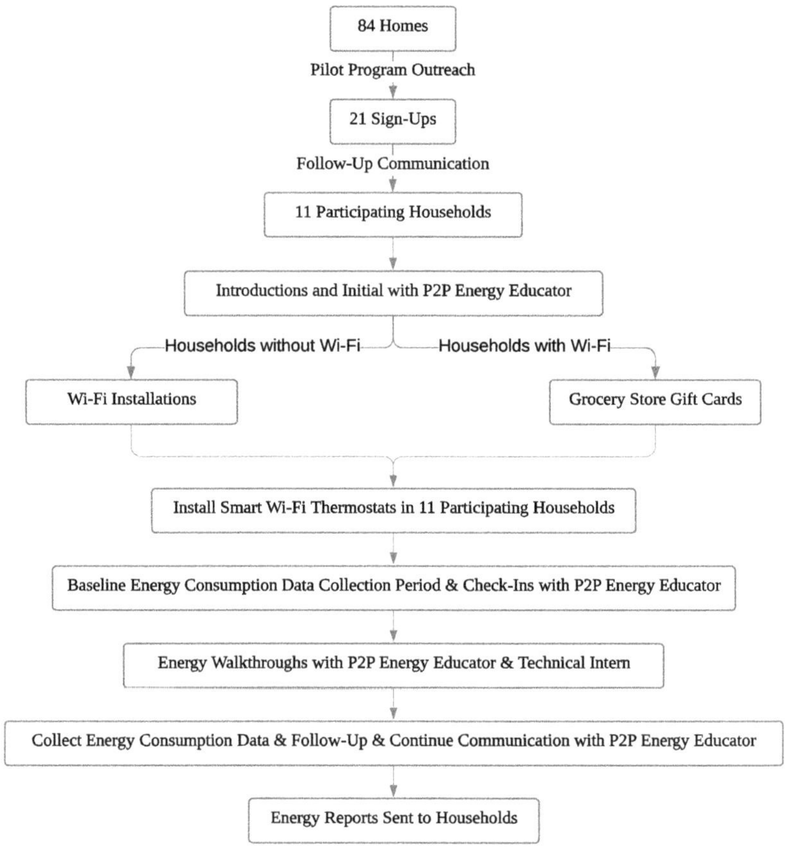

Figure 1. Pilot program process diagram: the pilot program began by conducting outreach to the 84 homes, which resulted in 21 initial sign-ups and 11 final participating households. Following initial introductions and communication with the P2P energy educator, Wi-Fi was installed in participating households that did not have Wi-Fi and grocery store gift cards were distributed to households with Wi-Fi already installed in order for smart Wi-Fi thermostats to be installed in the 11 participating households. There was then a short period in which baseline energy consumption data was collected when the P2P energy educator maintained frequent communication with participants. Energy walkthroughs were then completed with each household, and after further data collection and follow-up with the P2P energy educator, individualized energy reports were generated and sent to each participating household.

An energy walkthrough was then completed with each participating household after the baseline data collection period. In collaboration with the P2P energy educator, the technical intern prepared a checklist, informational handout, and energy consumption report, which were used as guides for the energy walkthrough. The checklist was composed of energy saving behaviors and practices categorized by room and type. It also included additional questions and points of discussion that were to be addressed during the energy walkthroughs. A comprehensive and condensed version of this checklist was created to serve as an informational handout for participants. To provide participants with insight into how their energy consumption compared to those in their community, a report was generated that documented each individual home's energy consumption as well as the maximum, minimum, and average energy consumed in their neighborhood. Ultimately, the goal of the energy walkthrough was to begin the energy education process by introducing ways to reduce energy consumption, helping participants become aware of energy consumption patterns, and gaining an understanding of each household's specific needs and capabilities.

The energy walkthrough was primarily led by the P2P energy educator with the technical intern present to answer technical questions and be an additional resource and reference for educating participants. During each walkthrough, the P2P energy educator went over the energy consumption report with the participants. The checklist was utilized to discuss their current energy consumption practices, issues or concerns they had about reducing energy, and to walk through the house with the participants identifying energy reduction practices in specific rooms and for specific tasks. Lastly, an informational handout was provided to be used as a reference for the individual(s) present during the walkthrough and for any additional members of the household.

The P2P energy educator followed a similar approach for each walkthrough but tailored the process as necessary to acknowledge specific needs and reactions of participants. For example, the heads of some households included their children in the walkthrough, thus the P2P energy educator and technical intern worked to engage the children during the visit. Following the energy walkthrough, the technical intern documented the interactions and discussions with each participating household. A critical element of this documentation was to take note of home repairs or issues that were of concern for the household or that were prohibiting a household from being able to adequately reduce their energy consumption.

Following the energy walkthrough, energy consumption data continued to be collected and analyzed for the participating households. To document changes and progress and provide the households with feedback, monthly energy reports were created. These reports presented monthly household and neighborhood energy and cost savings based upon the measured energy consumption and the weather-normalized energy consumption obtained through the machine learning model developed for each residence. The savings were then converted to metrics that would provide a better understanding of how the savings translate to everyday life. Some of these metrics included the equivalent number of phones charged, number of trees saved, gallons of gas, and number of meals based on the energy and cost savings. The energy reports also included a simple tip for additional ways residents could incorporate energy savings behaviors into their lives and homes.

Regular feedback was incorporated into the program as a means to further establish communication and relationships with participants, build community engagement, provide additional energy education, and encourage the process of energy behavior changes, as described by the transtheoretical model of behavior change [36]. The intent was to send energy reports to the participating households on a monthly basis. However, due to the timeline of the energy walkthroughs and logistical changes within the program, the energy reports were not consistently sent and discussed with participating households.

3.3. Program Status at Completion of Pilot Program

After energy reports were sent to participating households with feedback based upon their response and energy behavior changes following the energy walkthroughs, the pilot program was temporarily put on pause to re-evaluate and measure progress of the program. Additionally, this time was spent adapting to the unforeseeable restrictions from the COVID-19 pandemic. This allowed the program to be restructured and strengthened for a relaunch and more complete implementation of the program in the neighborhood. A new P2P energy educator was also hired during this time and completed training and preparation to work with the participating households. Currently, the energy reduction program is continuing to be implemented in the initial neighborhood.

4. Methods

This study takes a multi-method and interdisciplinary approach to assessing the processes and efficacy of the peer-to-peer pilot program. Methods of assessment include surveys of residents, collections of program notes, interviews with program leaders, and home energy use data tracking.

4.1. Participant Survey

Energy saving behaviors and behavior modifications are complex and dynamic in nature, and numerous studies and analyses unveil the psychological, physical, social, and situational facets of such complexities. However, limited research exists focusing on these topics solely within underserved and low-income communities. Surveys were, therefore, created and administered to participants in the pilot program to gain insight on participants' current energy usage trends, initial impressions of the program, and energy consumption values and motivations. Furthermore, survey data provides information on individual needs and interests of participants. This can be used to facilitate future interactions, tailor the program to particular households, and understand nuances on the views and realities of residential energy use within the neighborhood. According to Fredericks et al., 2015, environmentally conscious behavior and the ability to modify behavior is influenced by socio-demographics, situational factors, and phycological and personal values [37]. Thus, the survey was structured into three categories: demographics and general information, program experience, and values and motivations.

The portion of the survey evaluating the values and motivations of energy behavior was based on previous literature, which highlights the numerous factors that are associated with and influence behavior change and environmentally conscious actions. Consequently, values and motivations must be analyzed from a holistic viewpoint that does not isolate single factors but instead examines the interconnected nature of all factors. These findings, therefore, were used as a guideline for the types of questions and topics to include in the participant survey when investigating energy behavior and the connection to personal values and motivations.

The survey design was based upon a survey methodology employed by Carrus et al., 2008, in a study conducted to evaluate recycling and public transportation behaviors [38]. Because this study analyzed similar overarching topics associated with environmental and behavioral actions, the original questions were tailored using the above findings for the purpose of this study.

Generating the survey for participants not only required research into the content of the questions but also careful consideration for how the survey was structured. The survey included various types of questions such as rankings, agree or disagree, multiple choice, and free response. Each question was carefully analyzed to evaluate the question format to utilize, the proper language to use, and where to include the question within the survey. These considerations were taken to prevent discrepancies between participants' understanding of questions and to prevent responses from being influenced by the organization and framework of the survey.

Surveys were administered to participating households during the energy walkthrough with the P2P energy educator and technical intern. For completing the surveys, households were incentivized with a gift card to a local grocery store. In total, eight surveys were completed and analyzed for this study.

4.2. Interviews

To analyze the internal processes, experiences, and takeaways of the pilot program, interviews were conducted with key figures involved in the program's development, implementation, and advancement. All interviews were held remotely via video conference. Each interviewee was asked a series of the same general questions as well as individualized questions based upon the nature of the work they completed and their contributions to the program.

Interviews were conducted to examine the perspectives and experiences of internal sources from each angle of the program. With certain individuals working on the technical and program logistics and others working on community development directly with the residents, these interviews would determine nuances between the experiences and views of each individual and trends among their responses. The ultimate purpose was to inform which characteristics of the implemented program needed to be reevaluated and reanalyzed and to help envision needed processes for future advancement of the program in the local context associated with this research and elsewhere.

The individuals chosen to be interviewed for this study occupied different positions in the pilot program and were involved in various stages of its development, implementation, and advancement. A total of five individuals were interviewed for this study. This included the (1) P2P Energy Educator, (2) Technical Intern, (3) Nonprofit Director, (4) Program Innovator and Energy Analyst, and (5) Program Coordinator. Brief descriptions of each interviewee with their respective role and contributions in the program can be found in Table 2. Additionally, for the pilot program, the positions of the P2P energy educator and technical intern also included assisting with the development and logistics of implementing the program.

Table 2. Descriptions of Interviewees' Roles.

Position	Description (Roles, Responsibilities, and Contributions)
P2P Energy Educator	Responsible for interactions with residents and was the point of contact between the participating households and the rest of the program; served as a mentor and peer to the residents and provided education on energy saving behaviors and tools; responsible for relationship building with residents; established initial communication with households who signed up for the program; scheduled meetings for thermostat installations, energy walkthroughs, and all other interactions; assisted with thermostat installation; and maintained regular communication with residents to follow-up on meetings and address questions.
Technical Intern	Worked closely with the P2P energy educator interacting with households but with a greater focus and background on the technical aspects of energy savings; assisted with initial outreach and thermostat installations; created preparatory materials and documents for household interactions and energy walkthroughs; and kept track of technical related issues and concerns from interactions.
Nonprofit Director	Focused on determining and navigating the role of the energy reduction program within the overall purpose of the nonprofit; sought to create partnerships and make connections with other community organizations to further the work of the program; primary fund-raiser for the non-profit; and managed the budget.
Program Innovator and Energy Analyst	Collected historical energy data on residences, identifying the opportunity to realize behavior-based savings; responsible for the ideation of the program with the intent to build capacity within the neighborhood; introduced and proposed this program to the nonprofit director and was the primary figure in the development of the program and the early stages of partnership development; worked with utilities to collect energy data for participating residences; and responsible for the measurement of savings realized for each residence and collectively.
Program Coordinator	Community partner who worked for the nonprofit, managing numerous programs and initiatives; began working with the program near the end of implementation of the pilot program and transitioned into the role of overseeing the program; restructured the program and prepared for a new P2P energy educator after the initial pilot program; and focused on story development of the nonprofit and program to increase presence and awareness within the neighborhood.

4.3. Tracking Energy Consumption and Measurement of Savings

To measure energy consumption changes as a result of actions taken during the program, access to the monthly energy consumption data of residents was essential. All participants were first asked and agreed to sign release forms guaranteeing confidentiality and permitting researcher access to their monthly energy consumption data. Researchers then worked collaboratively with the energy retailer responsible for the monthly residential billing to receive monthly energy statements and energy consumption for the twelve months prior to initiating outreach for participating residents.

The energy characteristics of all houses in the study were obtained from reviews of the blueprints of all houses. Insulation values in metric units for the walls and ceilings were respectively RSI 3.17 and RSI 6.34. All windows were double-paned. The heating systems were all gas-condensing furnaces, with efficiencies of 0.95. All air conditioners were associated with SEER values of 14. The only difference between houses were the floor areas, with residences having one of two values (110 or 140 m^2).

Weather data for each energy consumption meter period, including the twelve months prior to beginning the study, was obtained from the NOAA's Climate Data Online resource [39]. Collectively this data was used to build baseline energy models for each residence to predict energy consumption for any energy consumption meter period given the weather conditions present during that meter period. The approach employed by Al Tarhuni et al. [40] and Alinezi et al. [41] was utilized to develop these models, which effectively permit prediction of the monthly energy consumption based upon resident energy use patterns present prior to the start of the program. Fundamentally, this approach synchronizes monthly energy consumption data with local outdoor weather data (obtained from the NOAA Climate Data Online resource [39]). For each meter period, average outdoor temperature and probability densities for the outdoor temperature falling within the 60 uniformly spaced temperature bins ranging from outdoor temperature bins ranging from −23.33 to 37.78 °C were used. A machine-learning algorithm based upon a stacked ensemble approach [41] was used to develop a model to predict monthly energy consumption for each residence based upon the energy consumption prior to the study. This model, when applied to post-program energy consumption data, enables prediction of the energy consumption based upon pre-program behaviors. The difference between the predicted consumption from the model and actual consumption for each meter period describes the change in consumption as a result of program actions.

Post initiation of the program, residence energy consumption changes for each meter period were estimated utilizing the energy model to predict energy consumption. This prediction was compared with the actual consumption. If the actual consumption was less than the predicted consumption, energy savings were inferred, whereas if this consumption was more than predicted, energy increases were inferred.

5. Results

5.1. Participant Survey

The survey results showed that out of the eight surveys completed, seven residents were aware of the Percentage of Income Payment Plan (PIPP) Program, an energy assistance program, two were enrolled in PIPP, and three were interested in learning more, as shown in Figure 2a. Additionally, when asked to indicate current energy saving behaviors, the surveys revealed that many households were aware of and engaged in energy saving behaviors in more than one way. This is revealed in Figure 2b in which lighting, heating and cooling, and washing and drying clothes all were marked by six or more households as ways they were already attempting to reduce energy consumption. Finally, Figure 2c shows that motivators for adjusting thermostats vary in importance but implies that personal and family comfort influences thermostat adjustments the greatest.

It is vital to note that due to the limited reach of the pilot program and small number of responses, generalizable conclusions cannot adequately be drawn from the presented results. Rather, the responses serve as a means to further understand the implementation, development, and evolution of the program and to consider needs and characteristics of the neighborhood and households that otherwise may not have been observed.

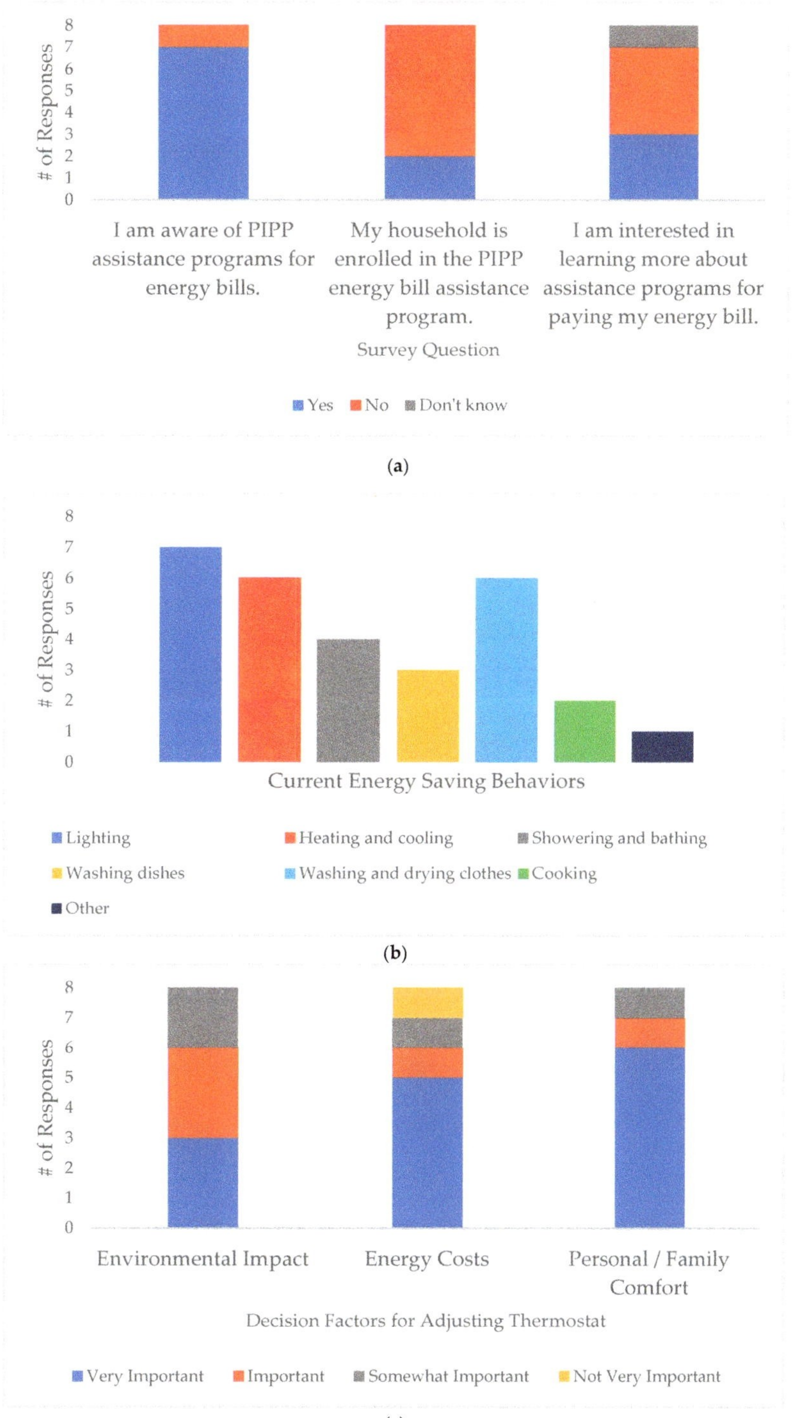

Figure 2. Participant survey results: (**a**) energy assistance programs; (**b**) energy behaviors; (**c**) thermostat motivators.

5.2. Interviews

Upon completion of interviews with program organizers and coordinators, an analysis was conducted to examine responses. Five key areas and trends of takeaways, recommendations, and insight into the future of the program were identified through this analysis: (1) community empowerment and sense of control, (2) education and training, (3) program impact and reach, (4) program requirements and logistics, and (5) program as integrative and collaborative.

5.2.1. Community Empowerment and Sense of Control

Each interviewee was asked to describe their experiences and views on the success of the program. Their responses were derived from the limited results and evolution of the program beyond the initial pilot program. A common theme among the responses highlighted that residents were able to acquire a new sense of control and empowerment. According to the P2P energy educator, a successful aspect of the program was *"having people understand they could take control of their utility bills by reading and understanding them."* By learning about the relationship between behaviors and utility bills, residents were able to see that they could take control of their utility bills. This not only increased their sense of control in their understanding of their utility bills but also showed them their role in addressing climate change as well as how they could take control of aspects of their lives beyond energy consumption in what seemed to be simple and trivial ways. As explained by the program coordinator, *"people understand they have more control in simple things in their lives than they think"* and many do not know they can save money based on their thermostat, which then translates into curiosity of how and where else they can save money. The pilot program empowered residents to see their actions and behaviors as a way to gain control of their utility bills and other areas of their lives. Thus, the lives of participating households were impacted beyond energy and the primary scope of the program.

5.2.2. Education and Training

The responses provided by the interviewees brought forth the crucial role of the education and training required by individuals working in the program and the areas in the pilot program where education and training needs were insufficiently met. From a general and program-wide outlook, more intentional training and knowledge was needed for individuals in the program, particularly for the P2P energy educator and technical intern. There were two primary areas in which further training was necessary. First, greater attention was needed on technical knowledge such as utilities, utilities bills, energy programs, and miscellaneous specifics on energy consumption and savings. Second, the P2P energy educator and technical intern expressed a lack of training and preparation on soft skills for their specific roles as well as the program as a whole. Such training and skills included communication and people skills necessary for working with individuals with different levels of technical expertise within the program and also the knowledge of how to properly and consciously communicate with community residents and understand appropriate language to use. Furthermore, the interviewee's responses indicated a goal to strengthen the opportunity to provide education to residents on utilities, utility bills, energy programs, etc., which is dependent upon the knowledge and education of those in the program.

With the P2P energy educator being the primary point of contact and the person in charge of conveying information and education materials to residents, there were specific details identified of what education and training is essential for this role. According to the P2P energy educator, they did not feel adequately prepared to confidently and comfortably work and interact with fellow program developers and with residents. They suggested greater collaboration and education from program directors to feel more confident working with those in technical roles such as engineers and energy analysts. While the P2P energy educator's role is to be a peer and mentor to residents in the program, it is vital for them to gain an in-depth understanding of the technical components of the program in order to increase self-confidence and amplify their impact when working with residents. As the P2P energy educator stated, *"They're counting on me to know something . . . I should have answers."* This includes the technical knowledge as previously indicated and knowledge of other community programs and organizations, both energy and non-energy related. Discussing energy with residents revealed insight into why certain households have high energy consumption, thus the P2P energy educator should be able to provide knowledgeable recommendations and assistance such as how to get mattresses or warm clothing if that is prohibiting a household's ability to turn down their thermostat and reduce energy consumption.

5.2.3. Program Impact and Reach

Another common trend among responses was an understanding of the impact the program had within individual households and within the community at large. A predominant takeaway was the need to include all household members in the energy education process. While conclusive energy consumption changes and savings were not able to be made based on the limited time frame of the program, as well as complications stemming from the COVID-19 pandemic, this understanding was significantly driven by the fact that the household whose energy consumption decreased substantially after the energy walkthrough had all household members present during the walkthrough.

This was further highlighted by the feedback the P2P energy educator and the technical intern received from residents during the energy walkthroughs. The feedback they received revealed that, while residents were receptive to a majority of the tips on ways to save energy in their homes, there were limitations and challenges in maximizing the impact and energy savings. For instance, the technical intern expressed that many residents appeared to believe they lacked the power and ability to save energy if they felt certain energy use was out of their control. Similar sentiments were echoed by the P2P energy educator, stating, *"The houses were not our variables . . . but in some ways, the houses were dramatically different . . . a lot was (blamed on) the insulation of the house (which actually was the same for all houses) or (their) kids for the usage."* These insights, therefore, reveal the necessity of incorporating the entire household in the energy education process as well as other obstacles homeowners face to realize the impact of their actions and barriers that inhibit the savings they are able to achieve.

To reach the entire household it was also clearly expressed that the approach must carefully consider how information is conveyed to the adults and children within households. According to the P2P energy educator, they were sometimes intimidated and concerned they would come across as arrogant when discussing energy savings tips with adults because energy savings is inherently tied to one's finances and, therefore, can be a sensitive topic. Ultimately, positively impacting a resident's energy savings behavior is complex, and best practices need to include entire households in the process and the aforementioned concerns that were revealed by the interviewees.

In addition to addressing the impact of the program on the household level, analyzing the impact on the community level was a critical factor that influenced the program's impact. As the program and energy education process began, it was quickly recognized that the time required to develop relationships and trust within the community is longer than initially anticipated. According to the nonprofit director, *"We learned early on that the approach is too simple. The idea that you could establish trust with a group, a new community interface for them, and begin to change their behavior quickly was an incorrect assumption."* Similarly, the program innovator and energy analyst said, *"I had actually seen this initiative as being an example of how to combat climate change nationally with speed . . . and I think the greatest learning and impact that I've had is that it is slow and about developing relationships over the long term."* It is evident that the program must first establish relationships and trust within the community to enable connections with households on an individual level, which takes time and must be a long-term endeavor. This must include not only relationships with community members but also with community organizations and programs already working in the community. Thereby, the program and P2P energy educator can leverage the community's assets to collaborate and work with the community to assist in meeting the needs of the community and households. It is critical that the program and all partners acknowledge the amount of time required to establish relationships and that they have the bandwidth to accomplish this.

Based upon the interview responses, a significant component of establishing relationships is improving the presence and familiarity of the program in the community, something that lacked in the initial implementation of the pilot program. According to the program innovator and energy analyst, *"I guess I just didn't initially realize it would just be so challenging to get people to sign up . . . . we realized that step one has to be the establishment of relationships with community members in order to potentially establish trust."* This response not only reiterated the aforementioned requisite for relationships and trust but provides further insight into gaining interest in the program from the community. The interviews revealed that the program needed greater community involvement and exposure from the beginning. It was suggested that in order to achieve this, the community must be part of this process to increase trust and familiarity and to ensure the program is driven by the community.

Lastly, the ability for this program to make a positive impact requires an in-depth understanding of the community. As indicated by the program coordinator, *"This work needs to build to much greater system change, to energy democracy, and to what it means to actually be in charge of your neighborhood and its health and vitality . . . It's pretty unfortunate how much we didn't know."* Through the interactions with the community, the reality of the systemic issues and unnecessary dependencies the community endured became apparent. Particularly, simple things that may not typically be questioned or considered must be part of the process. Thus, while maintaining the goal of reducing energy consumption, the program and those involved must have a broad and in-depth understanding of the perspectives and experiences of the community beyond energy and energy savings alone.

5.2.4. Program Requirements and Logistics

Mutual recommendations, perspectives, and critiques of the requirements and logistics of the program were also revealed through the interviews. A particular need that was identified was the need for reliable funding and financial support. According to the program innovator and energy analyst, *"We just realized that it is going to take time, and we've got to figure out a funding resource to help make that time feasible in the end."* Thus, as a deeper understanding of the length of time required to establish community relationships and trust was acquired, it became evident that a greater funding source would be needed to create long-term and lasting community presence.

One element of the program that requires funding is the incentives residents receive for their participation. However, it was revealed that if these incentives continue in the future, they must be more intentional. This was clearly articulated by the P2P energy educator and their interactions with residents *"I do believe that incentives work. I think that we could have done different things with the money that would have helped better if we were looking at it more individualized because we were looking at a broad spectrum... every single person that we're working with is dealing with different reasons why their bills are the way they are."* To meet the goals of the program and make the long-term presence financially viable, it is suggested that incentives be utilized in a more purposeful manner that further aligns with the needs of the program and the individual situations of the residents. Fundamentally, if financial incentives are offered, they must reward energy savings and serve individual household needs.

It was also revealed that a more detailed plan must be established for the implementation of the program. Based on the responses from the interviews, it is critical to have short-term and long-term plans that emphasize both the technical and conceptual elements of the program and that also are built on the understanding of what sustainability means to the community. According to the program coordinator, *"We need to know what it [sustainability] means to them based on their language and how they live on a day-to-day basis and adapt how we think it should be implemented in their neighborhood."* Therefore, the program must balance the focus on the program specific goals of energy savings and on taking a holistic approach of what is needed to achieve community sustainability and resilience beyond energy. This implies that boundaries must be set on how far the program, as well as the role of the P2P energy educator, can veer off focus. Ultimately, for households to achieve energy savings, there are additional factors necessary to be acknowledged and included in the work.

Furthermore, the interviewees discussed the necessity to consider all angles of the program, of the community, and of any potential issues that may arise before beginning the implementation process. In order to implement this, the program plans must not overlook simple characteristics and understandings of the community and require significant communication and collaboration with all individuals involved in the program and with community members. Finally, it was indicated that in order to meet these requirements and account for the details necessary for the implementation of the program, it is vital that plans established are adaptable and have the ability to evolve as new needs and understandings are discovered.

5.2.5. Program as Integrative and Collaborative

A final trend revealed through the interview responses is the necessity of the program being integrative and collaborative, which, while discussed in previous sections, deserves further emphasis and detail. To maximize community presence and build relationships and trust, feedback from the interviewees strongly suggested to not only establish partnerships with existing organizations already operating within the neighborhood, but to also implement this program into the work of an existing organization already trusted, rooted in, and represented by the community. By having the program under the umbrella of such an organization with well-established community presence and partnership, the time and work required to build new, long-term relationships and trust will be mitigated.

Other recommendations for taking a more collaborative approach included having the community identify a P2P energy educator, creating opportunities for youth and high school students to get involved, and having the outreach and presence of the program be completed by community members themselves. According to the program innovator and energy analyst, *"I would also encourage the community to identify a peer-to-peer person whom they would want to hire to manage the program, and we would actually manage the program through that organization … it would be transparent; they would be seen as the enablers of their community."* The initial belief was that the pilot program addressed the need of community engagement by having a P2P energy educator with similar experiences and by incorporating the program into the work of a nonprofit. However, these sentiments reveal the depth at which this must be implemented and indicate the recommendation of redirecting practices within the program to be driven and operated by the community itself.

Finally, insight from the individuals involved in the development and implementation of the program highlighted the potential for the program to serve as an opportunity for broader community development and work beyond energy savings. According to the program coordinator, *"It just starts the conversation for future work that is much bigger than just saving a few dollars in your home. Like, what is it going to be [to build] a truly sustainable and resilient self-sufficient neighborhood."* Thus, future programs should not focus on energy and energy cost savings alone but should, instead, integrate with other goals and needs of the community and community organizations. This was further emphasized by the idea that the program can provide the impetus to create greater system change and advance the efforts underway to achieve community resilience.

5.3. Preliminary Home Energy Usage Result

Comparisons of the household energy usage summed over the three month period following the energy walkthroughs and the monthly energy savings are shown in Figure 3a,b.

Out of the eight households included in the analysis, three showed energy savings and the remaining five showed increases in energy usage. For each month of energy data collected in 2020, the household averages were 760, 630, and 665 kWh respectively, slightly greater than averages for 2019 energy data. Analyzing the energy savings for each house during each month of available energy data, there was a maximum energy reduction of 53.17% and a maximum energy increase of 88.76%.

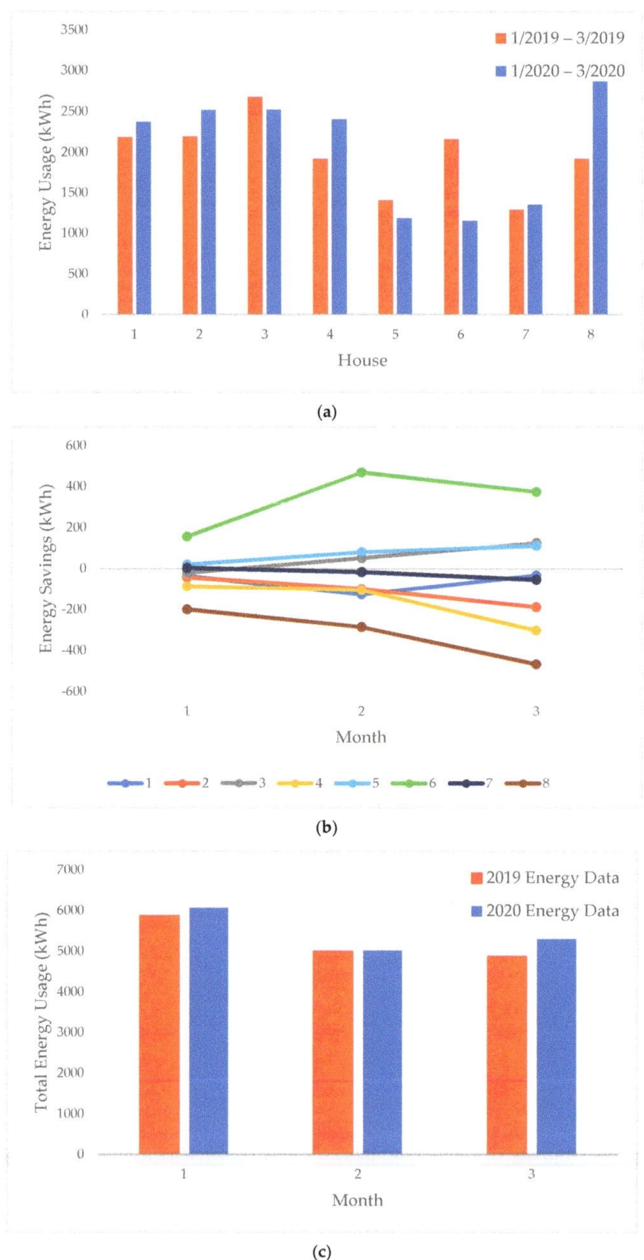

Figure 3. (a) Total household energy usage for 3-month period; (b) household energy savings (3-month period post-energy walkthrough); (c) total energy usage, 2019 and 2020 comparison.

Summing the energy usage of all participating households, energy changes between 2019 and 2020 were less appreciable, as shown in Figure 3c. Overall, the participating households experienced an energy increase of 4.11% over the three-month period from 2019 to 2020. However, this amount is within the uncertainty of predicting energy consumption and savings. This equated to a total energy increase of 648 kWh, equivalent to a cost of $76.40 and 0.46 metric tons of CO_2 emissions (Table 3). As evident in Figure 3a, one household, house 8, was an outlier in terms of the extent of energy increase with a total increase of nearly 950 kWh, negating all energy reduction achieved by other households. Furthermore, the aggregate energy consumption reveals the greatest energy increase was seen during the third and final month included in the analysis.

Table 3. Household savings (total for 3–month period).

House	Energy (kWh)	Cost ($) [1]	CO_2 (Metric Ton) [2]
1	−191	−$22.52	−0.14
2	−327	−$38.55	−0.23
3	156	$18.39	0.11
4	−489	−$57.65	−0.35
5	216	$25.47	0.15
6	1002	$118.14	0.71
7	−65	−$7.66	−0.05
8	−950	−$112.01	−0.67
Average	−81	−$9.55	−0.06
Total	−648	−$76.40	−0.46

[1] Energy costs of $0.1179/kWh for Dayton, Ohio [42]. [2] 7.07×10^{-4} metric tons CO_2/kWh [43].

6. Discussion and Improved Program Design

With participation of 11 out of a potential 84 homes, qualitative and conclusive outcomes cannot be made. Rather the presented results serve as a means to improve the design of an energy reduction program for low-income communities that utilizes peer-to-peer education and focuses on energy saving behaviors. In the improved program design, existing literature and anecdotal evidence from the pilot program are leveraged to emphasize the critical role of community engagement. The following discussion highlights key insights, considerations, and improvements on how this program can be strengthened and improved as it evolves and is scaled and expanded beyond the initial neighborhood.

6.1. Maximizing Program Impact

A primary takeaway from the interview responses, which was confirmed based on the results of the participant survey and energy data, was the need to improve the impact of the program on a community and household level. While the need to establish community relationships and trust was known to be a challenge and essential component of the program from the beginning, the time required and the steps necessary to achieve strong and impactful relationships was underestimated. Establishing a robust presence in the neighborhood is, thus, a vital component that should be the focus of the program before beginning energy education on a household level.

In addition, the importance of providing energy behavior education to all members within a household was revealed. This was shown to anecdotally influence the ability for households to achieve energy savings and is supported by the results of the participant surveys, interview responses, and energy data. Although the available data and responses are limited, household members indicated that the ability to modify energy use habits in their household was more easily attainable for themselves and more difficult for others within their household. However, the available data and responses are limited, and further research is needed. Moving forward, a goal of educating households on energy savings behaviors should emphasize collaboration among all household members in a manner that acknowledges the complexity of and accounts for the diversity among household compositions and roles of household members within low-income communities.

6.2. Understanding Program Reach

Preliminary results revealed gaps in the ability to accurately understand and analyze the available yet limited data. One significant challenge the program faced from the onset of the program was gaining interest within the community for households to sign-up and participate-limiting the reach of the program and ability to properly evaluate results.

With 11 participating households out of the 84 households in the neighborhood who were contacted and informed about the program, it was uncertain as to what appealed to those 11 households and what barriers existed that hindered greater interest. While mail-in flyers were sent to each household and additional door-to-door outreach was conducted in an effort to increase community presence and interest, the limited participation clearly demonstrated the lack of community engagement and need to gain community buy-in and involvement from the inception of the program. As previously discussed, strong presence and trust are indicators for gaining interest and engagement. Evaluation of survey and interview responses and energy data also revealed further insight into potential factors that should be explored further in the future. Based upon the participant survey results on utility bill costs and assistance programs, the indicated energy burdens endured by participating households were not as high as initially anticipated. This raises the question as to whether energy burdens are not exceptionally severe in the particular neighborhood or if among all the households in the neighborhood, those with less severe energy burdens were those who opted in to participate in the program. It is possible that households with the most severe energy burdens did not have the bandwidth to participate and that the challenges and realities of living in financial poverty limited participation. This question requires further attention in future research.

The reality and uncertainties of living in financial poverty revealed considerations necessary for properly analyzing energy consumption and understanding the capacity for households to modify energy behavior. The energy consumption data for the three-month period following the energy walkthrough in comparison with the same three-month period from the prior year was significantly lower for one household and significantly higher for another household (houses 6 and 8 shown in Figure 3a, respectively), and out of the households with available data, three reduced and five increased their aggregate energy consumption between the 2019 and 2020 three-month period. However, certain characteristics and situational occurrences must be incorporated into the analyzation process to accurately interpret the data for these and future results. This includes factors such as changes in the number of household members, significant lifestyle changes, and changes in employment and the accompanying work schedule.

Furthermore, comparing energy consumption before and after households begin energy education implies baseline energy consumption would be stable or typical for a household. The experiences and observations by the P2P energy educator, the technical intern, and other program contributors, however, revealed the inconsistency of living in financial poverty, which may result in inconsistent energy consumption within households based on both behavior and lifestyle. Additionally, several households do not and are not able to live in one home or neighborhood for long periods of time. This adds additional challenges for analyzing data and establishing trust and relationships with individuals, which will require an effective way to determine baseline energy consumption.

As previously discussed, the greatest increase in aggregate energy use among the eight households occurred during the third month following energy walkthroughs. This may indicate that implementing energy savings behaviors and modifying behaviors declines over time. However, the third month of data corresponded to the beginning of stay-at-home orders set in place in response to the COVID-19 pandemic. It is, therefore, evident that the impact of external factors and situations that are beyond the control of households, communities, and the program must be included and examined when analyzing energy consumption data.

Ultimately, this understanding indicates that educating households on energy behaviors and analyzing energy consumption data must account for the inconsistency certain households experience. To accurately incorporate and understand the inconsistency and unpredictability households in underserved communities endure, further research and community insight will be essential, which can include interviews with participating households from the pilot program, particularly with residents whose energy consumption decreased following the energy walkthroughs.

6.3. P2P Energy Educator

The P2P energy educator is a central feature of the energy reduction program that aims to provide energy savings education to households through comfortable and trusting relationships. The P2P energy educator from the pilot program understood the lifestyles of households they worked with based upon their own background, but they lacked the understanding of such experiences as an adult and as a member of the specific community. This inhibited their ability to fully connect with residents and feel confident in their role as a peer-educator and revealed that greater care must be taken when selecting an individual to fill the position of the P2P energy educator. Potential ways to address this concern include having the P2P energy educator be a resident from the community, seek out individuals who are already trusted and respected within the community, and have community members nominate and elect individuals for the position.

Before the P2P energy educator begins working with households, introductory preparation is crucial to ensure they feel confident and comfortable when interacting with participating household members. A more formalized and intentional on-boarding process is recommended based upon experiences from the first P2P energy educator and other program contributors. This process may include education and training on energy, utilities, and applicable residential programs, introductions to and meeting with local organizations, regular and consistent collaboration and communication with other program organizers, and attending community events and outreach.

Once the P2P energy educator begins interacting and meeting with individuals, it is important to establish a robust tracking and communication process. Creating a system in which the P2P energy educator is able to track and take notes of any barriers that may prevent households from being able to achieve energy savings and make energy behavior modifications will ensure there is consistency between visits and between households, that they are on top of requests and needs from particular visits, and that they are better able to tailor the program for households.

As the P2P energy educator position further evolves and develops, it will be beneficial to define all responsibilities in detail and set boundaries for the position. Energy and utility bills can be a personal subject matter for households because of its relation to money and financial security and is, therefore, an intersectional issue that brings an array of interconnected factors into the conversation of what impacts and influences a household's energy behaviors and energy consumption. There must be a boundary established to determine how far their work can expand beyond the focus of energy savings to ensure other needs are being addressed. This also includes distinguishing boundaries between the P2P energy educator and the residents to establish and maintain a trusting relationship while not going beyond their responsibilities in the program and staying within the lines of serving as a peer-educator. Clearly defining and understanding the responsibilities of the P2P energy educator is necessary for the individual themselves, other individuals working in the program, and residents. Based upon feedback and experiences from the pilot program, this will increase confidence in the P2P energy educator and their ability to have a greater impact on the program and lives of those participating.

6.4. Additional Recommendations

Based upon the results and outcomes of the pilot program presented in this study, there are additional recommendations and ideas for the future of the program and its framework beyond what has been previously addressed. First, the program must take a holistic approach to finding energy savings and assisting households modify energy behaviors. The use of incentives showed positive feedback and responses among residents. To address specific needs and burdens endured by households and what may prevent them from reducing energy consumption, intentional standards and practices for incentives should be established. To expand the work beyond energy behavior and energy savings alone, it is recommended that program coordinators carefully and methodologically establish a plan and defined boundary of what the program is capable of incorporating into their work. Finding this balance will require strong program management and organization that is established at the onset of the program. Such management must also be maintained to ensure the program sustains its mission while creating greater resilience within the community.

As this study reveals, community engagement and presence are key to its success and impact. It is highly recommended to collaborate and establish partnerships with existing community programs and organizations that have the capacity to contribute to the program's efforts or are able to amplify and support the program's presence in the community. These efforts should not only focus on incorporating the program into the community but also on incorporating the community into the program and making the program be driven by the community itself. For instance, a community focus and leadership group can be established to hear insight and perspectives from community members and to tailor the program to specific communities and their needs and aspirations for the program.

To engage residents of all ages, additional programs and processes could be instituted for younger community members. The technical intern can become an intern position to create employment opportunities and skills and knowledge training for young adults and youth in the community. Partnering with community programs provides the opportunity to incorporate energy savings education in pre-existing programming for youth in the community. Working with community partners also presents the possibility of creating a community art project or display to track and present energy savings for the community. This has the potential to amplify interest, engagement, and motivation through a visual display made by the community to highlight the collective impact of energy savings.

Finally, it is recommended to reconsider and modify the energy education approach. A potential option to explore is to begin the energy education process through community and group events in an effort for the community to acquire a greater understanding and trust of the program. Feedback from this study and insight from previous studies indicate benefits of providing a casual setting for community members to socialize while also being introduced to the program by utilizing a peer-to-peer approach. It is particularly valuable for the P2P energy educator to establish and strengthen relationships with the community and individual community members. From this setting, individuals would then be able to sign-up for one-on-one interactions and meetings with the P2P energy educator to individualize energy savings behaviors and make it feasible for their lifestyle. Working on a community level first is expected to decrease intimidation or discomfort felt by residents and the P2P energy educator, which can arise when discussing what can be personal and sensitive topics and working in residents' homes. This approach also provides the opportunity to expand access to education to individuals who may not feel comfortable or have the capacity to work with a P2P energy educator on a personal level.

These insights and recommendations for an improved program design were developed from the outcomes of the pilot program. However, there are evident limitations of the current study that require further discussion. With only 11 participating households and the limited data available from household energy consumption and survey responses, the quantitative results that were collected do not provide statistical value. Rather, the results of this study are the recommendations for the improved program design, which were informed by the energy consumption data and survey results. Furthermore, it is critical to note that these results are also based on all 84 residences; notably, the significant variations in energy consumption among these similarly constructed houses reveals the potential in energy cost savings through behavior modifications and the lack of response, interest, and commitment to participate in the program provides value in understanding the necessity of community involvement and engagement. To investigate the efficacy of more intentional and intensive community input and engagement and, accordingly, the extent to which energy savings can be achieved through behavior modifications and peer-to-peer education in underserved communities, further research is needed.

7. Conclusions

Achieving residential energy savings through energy behavior modifications and a peer-to-peer education methodology in underserved communities is a complex and dynamic process, as presented in this study. It is evident that such a process requires robust community relationships that must be consistent and long-term. Because the timeframe necessary for establishing such relationships is beyond the scope of this study and the accompanying limited quantity of data, conclusive results cannot properly and effectively be drawn. However, the feedback, outcomes, and preliminary results presented provide insight into methods that contributed to the successes and drawbacks of the pilot program as well as recommendations to strengthen and scale the structure of the program.

This study indicates that a peer education approach is beneficial for gaining a genuine and individualized understanding of household barriers that exacerbate energy burdens. Areas identified as ways to increase the impact of the program include taking a holistic approach while maintaining the mission of the program, expanding the reach of the program on a household and community level, establishing a detailed and intentional long-term and short-term plan for implementation, and incorporating the community into the program itself. Further research and studies will be necessary, however, to determine the impact of the program and effectiveness of the preliminary feedback, results, and takeaways long-term. To improve community participation, and, thus, increase quantitative data, it is recommended that future programs begin with community level, grassroots efforts when conducting outreach to increase the number of participants, and focus on gaining in-depth insight from participating residents, such as by conducting ethnographic, qualitative interviews. Ultimately, this study demonstrates that the framework of a peer-led energy reduction program has the potential to not only reduce household utility bills but, if properly implemented, to contribute to the development of sustainable, resilient, and empowered communities.

Future research that utilizes the improved program design insights to scale and implement the program beyond the initial neighborhood and, ultimately, beyond the US is recommended. By leveraging global partnerships within the academic and research field that have a robust and pre-existing community presence, especially in the underdeveloped world, the reality of energy poverty and the capabilities for achieving energy savings through behavior modifications from the perspective of individual households and communities can be obtained. This has significant potential to expand existing understandings of global energy poverty and uncover nuances on energy behavior, behavior change, and community engagement across cultures on a local and global scale.

Author Contributions: Conceptualization, K.H.; methodology, J.H., A.G.R., and K.H.; software, K.H.; formal analysis, J.H., A.G.R., and K.H.; investigation, J.H., A.G.R., S.R., C.F., K.H., C.O., and B.P.; data curation, J.H., A.G.R., S.R., and K.H.; writing—original draft preparation, J.H., A.G.R., and K.H.; writing—review and editing, S.R., C.F., C.O., and B.P.; supervision, S.R. and B.P.; project administration, K.H.; funding acquisition, K.H. and B.P. All authors have read and agreed to the published version of the manuscript.

Funding: This research was funded through grants from the Ohio Housing Finance Agency's (OHFA) Technical Assistance Grant and the Marianist Foundation. The views and opinions expressed in this paper do not necessarily reflect the official policy or position of OHFA.

Institutional Review Board Statement: All subjects gave their informed consent for inclusion before they participated in the study. The study was conducted in accordance with the Declaration of Helsinki, and the protocol was approved by the Institutional Review Board of the University of Dayton on 28 June 2019 (45 CFR 46.110 Category (7)).

Conflicts of Interest: The authors declare no conflict of interest.

References

1. Special Report: Global Warming of 1.5 °C—Summary for Policymakers. Intergovernmental Panel on Climate Change. Available online: https://www.ipcc.ch/sr15/chapter/spm/ (accessed on 2 March 2021).
2. U.S. Energy-Related Carbon Dioxide Emissions, 2019. Energy Information Administration. Available online: https://www.eia.gov/environment/emissions/carbon/ (accessed on 5 November 2020).
3. Pellegrini-Masini, G.; Pirni, A.; Maran, S. Energy justice revisited: A critical review on the philosophical and political origins of equality. *Energy Res. Soc. Sci.* **2020**, *59*, 101310:1–101310:7. [CrossRef]
4. Lewis, J.; Hernández, D.; Geronimus, A.T. Energy efficiency as energy justice: Addressing racial inequities through investments in people and places. *Energy Effic.* **2020**, *13*, 419–432. [CrossRef] [PubMed]
5. Sovacool, B.K.; Dworkin, M.H. Energy justice: Conceptual insights and practical applications. *Appl. Energy* **2015**, *142*, 435–444. [CrossRef]
6. Jenkins, K.; McCauley, D.; Heffron, R.; Stephan, H.; Rehner, R. Energy justice: A conceptual review. *Energy Res. Soc. Sci.* **2016**, *11*, 174–182. [CrossRef]
7. Simcock, N.; Frankowski, J.; Bouzarovski, S. Rendered invisible: Institutional misrecognition and the reproduction of energy poverty. *Geoforum* **2021**, *124*, 1–9. [CrossRef]
8. Bouzarovski, S.; Petrova, S. A global perspective on domestic energy deprivation: Overcoming the energy poverty–fuel poverty binary. *Energy Res. Soc. Sci.* **2015**, *10*, 31–41. [CrossRef]
9. Residential Energy Consumption Survey (RECS): 2015 RECS Survey Data. Energy Information Administration. Available online: https://www.eia.gov/consumption/residential/data/2015/ (accessed on 2 March 2021).
10. Bednar, D.J.; Reames, T.G. Recognition of and response to energy poverty in the United States. *Nat. Energy* **2020**, *5*, 432–439. [CrossRef]
11. Drehobl, A.; Ross, L.; Roxana, A. How High Are Household Energy Burdens? American Council for an Energy-Efficient Economy. Available online: https://www.aceee.org/sites/default/files/pdfs/u2006.pdf (accessed on 2 March 2021).
12. Getting Energy Efficiency to the People Who Need It Most. Governing: The Future of States and Localities. Available online: https://www.governing.com/gov-institute/voices/col-cities-energy-efficiency-low-moderate-income-households.html (accessed on 21 February 2021).
13. Ouyang, J.; Hokao, K. Energy-saving potential by improving occupants' behavior in urban residential sector in Hangzhou City, China. *Energy Build.* **2009**, *41*, 711–720. [CrossRef]
14. Nguyen, C.P.; Su, T.D.; Bui, T.D.; Dang, V.T.B.; Nguyen, B.Q. Financial development and energy poverty: Global evidence. *Environ. Sci. Pollut. Res.* **2021**, *28*, 35188–35225. [CrossRef]
15. World Energy Outlook 2018. International Energy Agency. 2018. Available online: https://iea.blob.core.windows.net/assets/77ecf96c-5f4b-4d0d-9d93-d81b938217cb/World_Energy_Outlook_2018.pdf (accessed on 22 July 2021).
16. Analysis of the Voluntary National Reviews Relating to Sustainable Development Goal 7: 2018. United Nations. 2018. Available online: https://sustainabledevelopment.un.org/content/documents/258321159DESASDG7_VNR_Analysis2018_final.pdf (accessed on 22 July 2021).
17. Energy Access Outlook 2017: From Poverty to Prosperity. International Energy Agency. 2017. Available online: https://iea.blob.core.windows.net/assets/9a67c2fc-b605-4994-8eb5-29a0ac219499/WEO2017SpecialReport_EnergyAccessOutlook.pdf (accessed on 22 July 2021).
18. Jessel, S.; Sawyer, S.; Hernández, D. Energy, Poverty, and Health in Climate Change: A Comprehensive Review of an Emerging Literature. *Front. Public Health* **2019**, *7*, 357:1–357:19. [CrossRef]
19. Office of Community Services: LIHEAP Fact Sheet. Administration for Children & Families: U.S. Department of Health & Human Services. Available online: https://www.acf.hhs.gov/ocs/resource/liheap-fact-sheet-0 (accessed on 2 March 2021).
20. Weatherization Assistance Program. Office of Energy Efficiency & Renewable Energy. Available online: https://www.energy.gov/eere/wap/weatherization-assistance-program (accessed on 2 March 2021).
21. Roth, J.; Hall, N. Assessment of LIHEAP and WAP Program Participation and the Effects on Wisconsin's Low-Income Population: An Examination of Program Effects on Arrearage Levels and Payment Patterns. *ACEEE Summer Study Energy Effic. Build.* **2006**, 7-226–7-238. Available online: https://www.aceee.org/files/proceedings/2006/data/papers/SS06_Panel7_Paper19.pdf (accessed on 23 February 2021).
22. Lindenberg, S.; Steg, L. Normative, Gain and Hedonic Goal Frames Guiding Environmental Behavior. *J. Soc. Issues* **2007**, *63*, 117–137. [CrossRef]
23. Hines, J.M.; Hungerford, H.R.; Tomera, A.N. Analysis and Synthesis of Research on Responsible Environmental Behavior: A Meta-Analysis. *J. Environ. Educ.* **1987**, *18*, 1–8. [CrossRef]

24. Poortinga, W.; Steg, L.; Vlek, C. Values, Environmental Concern, and Environmental Behavior. *Environ. Behav.* **2004**, *36*, 70–93. [CrossRef]
25. Thaler, R.; Sunstein, C. *Nudge: Improving Decisions about Health, Wealth, and Happiness*; Yale University Press: New Haven, CT, USA, 2008.
26. Grilli, G.; Curtis, J. Encouraging pro-environmental behaviours: A review of methods and approaches. *Renew. Sustain. Energy Rev.* **2021**, *135*, 110039:1–110039:14. [CrossRef]
27. Establishing a Peer Education Program. The University of Kansas: Community Tool Box. Available online: https://ctb.ku.edu/en/table-of-contents/implement/improving-services/peer-education/main (accessed on 21 February 2021).
28. Ten Cate, O.; Durning, S. Dimensions and psychology of peer teaching in medical education. *Med. Teach.* **2007**, *29*, 546–552. [CrossRef]
29. Goldman, M.L.; Ghorob, A.; Hessler, D.; Yamamoto, R.; Thom, D.H.; Bodenheimer, T. Are Low-Income Peer Health Coaches Able to Master and Utilize Evidence-Based Health Coaching? *Ann. Fam. Med.* **2015**, *13*, S36–S41. [CrossRef] [PubMed]
30. Thorn, D.H.; Ghorob, A.; Hessler, D.; De Vore, D.; Chen, E.; Bodenheimer, T.A. Impact of Peer Health Coaching on Glycemic Control in Low-Income Patients with Diabetes: A Randomized Controlled Trial. *Ann. Fam. Med.* **2013**, *11*, 137–144.
31. Marshak, H.H.; de Silva, P.; Silberstein, J. Evaluation of a Peer-Taught Nutrition Education Program for Low-Income Parents. *J. Nutr. Educ.* **1998**, *30*, 314–322. [CrossRef]
32. Canuso, R. Low-Income Pregnant Mothers' Experiences of a Peer-Professional Social Support Intervention. *J. Community Health Nurs.* **2003**, *20*, 37–49. [CrossRef] [PubMed]
33. Twin Towers Neighborhood in Dayton, Ohio (OH), 45410 Detailed Profile. City-Data. Available online: https://www.city-data.com/neighborhood/Twin-Towers-Dayton-OH.html (accessed on 19 July 2021).
34. Prototype—Twin Towers, East Dayton, Ohio. CleanEnergy4All. Available online: https://cleanenergy4all.org/project/twin-towers-neighborhood/ (accessed on 5 November 2020).
35. Housing and Economic Development. East End Community Services. Available online: https://www.east-end.org/housing-and-economic-development (accessed on 5 November 2020).
36. Noar, S. Transtheoretical Model and Stages of Change in Health and Risk Messaging. Oxford Research Encyclopedia of Communication. 2017. Available online: https://oxfordre.com/communication/view/10.1093/acrefore/9780190228613.001.0001/acrefore-9780190228613-e-324 (accessed on 3 March 2021).
37. Frederiks, E.R.; Stenner, K.; Hobman, E.V. The Socio-Demographic and Psychological Predictors of Residential Energy Consumption: A Comprehensive Review. *Energies* **2015**, *8*, 573–609. [CrossRef]
38. Carrus, G.; Passafaro, P.; Bonnes, M. Emotions, habits and rational choices in ecological behaviours: The case of recycling and use of public transportation. *J. Environ. Psychol.* **2008**, *28*, 51–62. [CrossRef]
39. Climate Data Online. National Centers for Environmental Information. Available online: https://www.ncdc.noaa.gov/cdo-web/ (accessed on 23 February 2021).
40. Al Tarhuni, B.; Naji, A.; Broderick, P.G.; Hallinan, K.P.; Brecha, R.J. Large Scale Residential Energy Efficiency Prioritization Enabled by Machine Learning. *Energy Effic.* **2019**, *12*, 2055–2078. [CrossRef]
41. Alanezi, A.; Hallinan, K.P.; Huang, K. Automated Residential Energy Audits Using a SmartWiFi Thermostat-Enabled Data Mining Approach. *Energies* **2021**, *14*, 2500. [CrossRef]
42. National Renewable Energy Laboratory. Dayton Electricity Rates. Electricity Local. Available online: https://www.electricitylocal.com/states/ohio/dayton/#ref (accessed on 5 November 2020).
43. Greenhouse Gases Equivalencies Calculator—Calculations and References. Environmental Protection Agency. Available online: https://www.epa.gov/energy/greenhouse-gases-equivalencies-calculator-calculations-and-references (accessed on 2 March 2021).

Article

Just Transition as a Tool for Preventing Energy Poverty among Women in Mining Areas—A Case Study of the Silesia Region, Poland

Olga Janikowska [1,*] and Joanna Kulczycka [2]

1. Mineral and Energy Economy Research Institute of the Polish Academy of Sciences, ul. J. Wybickiego 7A, 31-261 Kraków, Poland
2. Faculty of Management, AGH University of Science and Technology, Mickiewicza 30, 30-059 Cracow, Poland; kulczycka@agh.edu.pl
* Correspondence: olgajan@min-pan.krakow.pl

Abstract: The inevitable energy transformation can be perceived as an opportunity and as a threat to the actions undertaken to prevent energy poverty in European mining regions. Silesia is a special exemplification of the European region whose economy has been based on coal industry for centuries. There are still about 70,000 miners and coal is also widely used for heating households. Based on developed map of jobs lost in mining and related industry and the demographic and social data the proposal of activities addressed to different group of people has been created. It was also indicated that energy poverty in Poland mainly concerns households inhabited by single women. Therefore, the major conclusion of the paper is postulated that the Just Transition strategy should be extended by issues strictly related to the situation of women in the future labor market. Additionally, the concept of a special hub for women, whose aim would be professional activation of women of various age groups, has been introduced.

Keywords: coal; energy poverty; just transition; women; sustainable development goal; gender; employment

Citation: Janikowska, O.; Kulczycka, J. Just Transition as a Tool for Preventing Energy Poverty among Women in Mining Areas—A Case Study of the Silesia Region, Poland. *Energies* **2021**, *14*, 3372. https://doi.org/10.3390/en14123372

Academic Editor: Ben McLellan

Received: 29 April 2021
Accepted: 1 June 2021
Published: 8 June 2021

Publisher's Note: MDPI stays neutral with regard to jurisdictional claims in published maps and institutional affiliations.

1. Introduction

Energy poverty has been identified in both developing and developed countries. It should be underlined that if energy poverty is defined as limitation of choice possibilities, it indicates that those who suffer this kind of poverty are a priori excluded not only from the possibility of the fulfilment of basic needs such as home heating or cooling and home cooking [1–5]. What is even more disturbing, they are excluded from other important elements that are necessary for individual development [6,7]. These include, above all, health, education, access to information, and participation in politics [8–10]. Energy poverty can have impact on heating or cooling a home, and also on health, education, access to information, and participation in politics. Energy poverty is a contributing factor in other types of poverty, and it can be perceived as one of the major challenges of the 21st century. It needs to be expected that in conditions of the absence of, or insufficient public policy support for those who suffer energy poverty [11–13], the level of energy poverty in Poland, especially in the Silesian Region, will increase as a result of the proposed legal regulations and the resulting requirements. Changes in the mining industry will effect on the employment structure in Silesia; additionally, in the near future, there will be a ban on the use of coal for heating households. Although it would affect all those who suffer energy poverty, it should be underlined that four times more households in energy poverty in Poland are occupied by single women than by a full family [2]. According to estimates the highest percentage of energy poverty in Poland is detected in the South-East part of the country (21–29%). The lowest is in the North but also in two mining regions: Śląskie and

Dolnośląskie voivodships (8–12%). In the near future, energy poverty in Poland, and in particular in the Silesia Region, will be influenced by two key factors. Firstly, transition to a decarbonized economy as 74,500 people work in the hard coal mining sector. Secondly, the Energy Performance of Buildings Directive, which proposes that the construction sector should not emit any carbon dioxide by 2050, but rather be carbon neutral. Additionally, all buildings in the European Union are to be decarbonized by 2050.

According to the Sustainable Development Goal (SDG) number 5: Gender equality, it is especially important to support and empower women and girls, as it would result in economic growth. What should be strongly underlined, is that although more and more women are present on the labor market, still in some regions women are deprived equal working rights. In this paper we are focusing on Just Transition, which is defined as transition of the workforce and the creation of decent work and quality jobs in accordance with nationally defined development priorities [14], and we claim that it should be a tool for preventing energy poverty among women. The Silesian Region was chosen due to its specificity in the role and place of women. For many years men in the region were professionally active and the miner's "honor" prevented his wife from being employed [15]. Additionally, the availability of jobs almost exclusively in the mining industry permanently consolidated the cultural division of social roles, at the same time the educational level of Silesia's women was rather poor. These historical premises, at some point, are present in the cultural code of Silesia. There is a concern that the region's profound economic transformation might result in structural unemployment, and, thus, deterioration of the material situation of Silesian women, which will result in energy poverty. As there is a clear relationship between access to energy and poverty, a more detailed analysis and prognosis are needed for people who live in extreme poverty, especially for women. It is very important to ensure proper financial and social support, which is foreseen with Just Transition and other restructuring funds [16].

In this article, we try to examine what changes will take place in the near future on the Silesian labor market and how they will affect the situation of women and energy poverty. We postulate that Just Transition should be a tool for preventing energy poverty among women in mining areas, and that a tailored program should be dedicated to women working in mining (about 10%) as a significant transformation is foreseen which will phase out coal mining in Poland by 2049.

2. Materials and Methods

Firstly, the search was conducted through both the commonly used technique of web research of literature and the review of academic interdisciplinary databases including Scopus and Google Scholar. The key words used in the search were "energy poverty", "just transition", and "women". It was shown that one of the most common approaches to defining energy poverty is understating the term as lack of access to a modern energy service, such as a situation where a household cannot afford an energy service or the energy itself to provide basic daily needs such as: heating or cooling, cooking, lighting [3–5,15]. Additionally, energy poverty can be seen as the limit in accessing environmentally friendly, good quality, safe, and affordable energy services [6]. The research pointed out that there is a clear relationship between access to energy and poverty. Therefore, the next step was to look at statistical data to identify the maximum number of coal workers employed in the EU and to compare it with future green job opportunities, with a focused on future job possibilities for women. The next part of the research concerned the energy restructuring in Poland. It was shown that the role of hard coal in the Polish energy sector will gradually decrease. Taking into consideration the multiplier effect in research, a map of jobs lost to mining was divided into areas closer to the mines and the more distant surrounding areas. The last part of the research is focusing on the Silesian Region, therefore, the next step was to look at statistical data with a special focus on the potential for creating new jobs and the role of women on the future labor market. The methodology adopted in this work is presented in Figure 1.

Figure 1. Algorithm for recognizing the market potential for Silesian women developed in this study.

During the critical analysis of literature, the research gap on the role and place of woman during the job restructuration in Silesia Region was identify. In the paper we suggest that energy policy is top-down policy whereas Just Transition should be a bottom-up policy, which means that the transformation processes should engage civil society and social capital. The paper represents a new approach to the Just Transition issues in the Silesia region. Analyze of the development of the national policy indicates, that since the very beginning to of the mining restructuration in Poland, there was a lack of reskill programs dedicated for a working at the mine industry women. During the research correlation between transition to a low carbon economy and energy poverty of women in Silesia region was indicated. During the transformation many jobs will be lost in the region, but still as we point out the social group who suffers the energy poverty in Poland are mostly single women. Due to a specific cultural tradition of the region married women usually stop they professional development to be a full time housewife's. During the research we have identify the knowledge gap, as aspects of the future situation of women on the labor market in Poland in relationship with the energy poverty has not been solved by existing studies. This is why we hope this paper would open the discussion on future place of women on the Silesian labor market after the energy transformation.

3. Results

Energy poverty is a phenomenon that occurs among the inhabitants of developed and developing countries, therefore, combating it is important in terms of social welfare [17]. With increased concern about climate change, analyses focusing on energy consumption, energy structure, and the energy market lie behind many policies. Energy poverty is related more to income poverty than to socio-economic inequalities [18–21]. In the EU, detailed analysis and assessment has been provided annually. The newest EU document on energy prices and costs in Europe shows that the annual increase in the price of electricity in the European Union has increased by 2.3% since 2010. This means that, taking into account inflation over the last decade, payable electricity is 9% more expensive. Moreover, in 2018 the poorest European households still spent 8.3% of their total expenditure on energy. Individual consumers who will take a part in transformation, on one hand would be protected from the growing energy prices, but on the other they are going to be fortified to become an active the energy market. However, even though the price of many renewable energy installations has dropped significantly, and more funds are available to support individual consumers, there are many individual houses, especially in Poland, where coal is widely used for heating.

The coal dominates the power sector in Poland, however with significant dependence on external supplies of natural gas and almost full dependence on external supplies of crude oil, the aspects of energy safety, transformation, and clean air have been usually the most important part of Polish energy policy for decades. The most important documents are as follows:

- 1997 The act of 10 April;
- 1997 The Energy Law (it has been often updated in last years as the consequence of the amendment of other legal in scope adopted primarily due to the need to adjust the Polish legal order to EU law requirements);
- 1990 Foundation of the energy policy of Poland for 1990–2010;
- 1995 Establishment of an energy policy until 2010;
- 2000 Assumptions of Poland's energy policy until 2020;
- 2002 Evaluation of the implementation and correction of the assumptions of energy policy until 2020;
- 2005 Energy policy of Poland until 2025;
- 2009 Energy policy of Poland until 2030;
- 2021 Energy policy of Poland until 2040 [22–25].

Poland's energy policy until 2040 defines the framework for energy transformation and is a response to the EU climate policy. The policy takes into account the scale of the challenges related to the adaptation of the national economy to the EU regulatory conditions related to the 2030 climate and energy goals, the European Green Deal, the COVID economic recovery plan and the pursuit of climate neutrality in line with national opportunities, as contribution to the implementation of the Paris Agreement. The energy policy in Poland was updated on the regular base, still up till now there is a lack of social emphasis in the policy, as it focuses mainly on energy safety, transformation towards RES and minimalization of the environmental impact. This is why we claim that Just Transition calls for in printing social challenges into energy policy. The core of energy policy of Poland until 2040 are illustrated in Figure 2.

With the new Energy Policy 2040 [22], approved in February 2021, a diminishing share of coal is foreseen in energy generation and heating. The main goals of policy is: ensuring energy security, safeguarding energy poverty, providing support for environmental protection, ensuring the safety of environmental controls, ensuring the safety of environmental controls, ensuring the safety of environmental controls, ensuring the safety of environmental controls, ensuring the safety of environmental controls, ensuring the safety of environmental controls, ensuring the safety of environmental controls, ensuring the safety of environmental controls. This policy is in line with the assumptions of the legislative package entitled 'Clean Energy for All Europeans' [26]. According to the European Commission, energy poverty cannot be capture only by a one indicator. Energy poverty is a phenomena which is correlated with following negative consequences such as; health and mental problems. As there is a lack of standardized definition of energy poverty, that is why member states should develop their own criteria according to their national context. What needs to be underlined in 2018 is that 6.8% of Europeans living in the private households could not keep up with their energy bills. The fact that energy and fuel poverty is still one of the biggest challenges of the 21st century is also widely discussed in the literature [1–6,9–14]. It covers technological, economic, and social aspects and also natural inequalities [27] which underline the fact that, still despite of the technological and scientific development of the world, quality of life does not follow. Therefore, energy poverty is identified both in developing countries, and developed countries. It should be carefully monitored in Poland as significant transformation is foreseen due to the phasing out of coal mining by 2049.

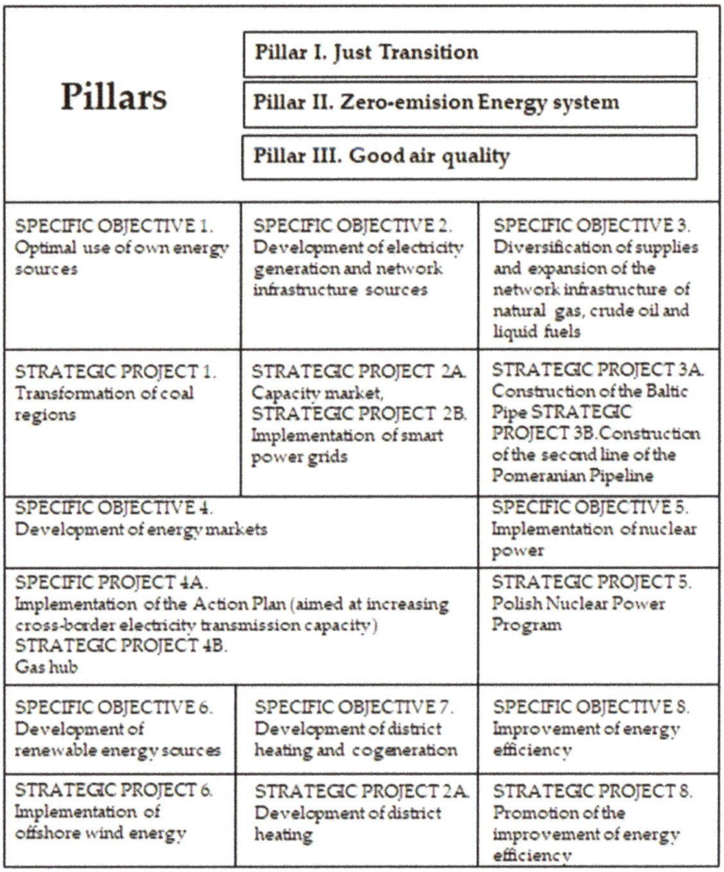

Pillars	Pillar I. Just Transition		
	Pillar II. Zero-emision Energy system		
	Pillar III. Good air quality		
SPECIFIC OBJECTIVE 1. Optimal use of own energy sources	SPECIFIC OBJECTIVE 2. Development of electricity generation and network infrastructure sources	SPECIFIC OBJECTIVE 3. Diversification of supplies and expansion of the network infrastructure of natural gas, crude oil and liquid fuels	
STRATEGIC PROJECT 1. Transformation of coal regions	STRATEGIC PROJECT 2A. Capacity market, STRATEGIC PROJECT 2B. Implementation of smart power grids	STRATEGIC PROJECT 3A. Construction of the Baltic Pipe STRATEGIC PROJECT 3B.Construction of the second line of the Pomeranian Pipeline	
SPECIFIC OBJECTIVE 4. Development of energy markets		SPECIFIC OBJECTIVE 5. Implementation of nuclear power	
SPECIFIC PROJECT 4A. Implementation of the Action Plan (aimed at increasing cross-border electricity transmission capacity) STRATEGIC PROJECT 4B. Gas hub		STRATEGIC PROJECT 5. Polish Nuclear Power Program	
SPECIFIC OBJECTIVE 6. Development of renewable energy sources	SPECIFIC OBJECTIVE 7. Development of district heating and cogeneration	SPECIFIC OBJECTIVE 8. Improvement of energy efficiency	
STRATEGIC PROJECT 6. Implementation of offshore wind energy	STRATEGIC PROJECT 2A. Development of district heating	STRATEGIC PROJECT 8. Promotion of the improvement of energy efficiency	

Figure 2. The core of energy policy of Poland until 2040, based on energy policy of Poland until 2040. Source: Energy policy of Poland until 2040.

Most documents emphasize that the ending of hard coal production in individual mines will relate to the provision of different social guarantees to workers, but that such a transition could also have a significant impact on many households using coal for heating. As there is a clear relationship between access to energy and poverty, there is a need for a more detailed analysis of people who live in extreme poverty and accompanying recommendations considering structural changes associated with the closing of the mining sector and the identification of people's needs and choices. According to Sen [27], "development consists of the removal of various types of unfreedoms that leave people with little choice and little opportunity of exercising their reasoned agency". He claims that the enhancement of freedoms allows people to lead lives that they have reason to value. According to his theory, the emphasis that has been put on gross domestic product (GDP) as the basic measure of development has been inadequate, and that is why his theory of development goes far beyond understanding development just as maximizing income per capita. He argues that measures of development should be followed by factors such as political freedoms, high standard of living, ability to access social goods, such as health care, education, reduction in hunger and premature death. Those are the basic conditions of development. Sen's contribution to the theory of development was a factor in the development of the Human Development Index (HDI), which is composed of the following indicators: life expectancy, education, and income per capita.

There is a strong correlation between energy poverty defined as the absence of sufficient choices and Sen's development theory as the core of development theory is not being excluded from those options that enable one to choose and obtain welfare in its broadest sense. If energy poverty is defined as the absence of sufficient choice, it indicates that those who suffer this special kind of poverty are a priori not only excluded from the possibility of fulfilling basic needs such as home heating or cooling and home cooking but, and this is even more disturbing, they are also excluded from other important elements that are necessary for individual development, and these are above all health, education, access to information, and participation in politics. In this context energy poverty is an obstacle to the development process, understood in line with Sen's theory as expanding people's freedom. Additionally, energy poverty reinforces development disproportions, which is especially important in terms of global justice theory [28,29] which is becoming a more and more important principle of political agendas.

The first difference revealed is a different distribution of the number of households ordered according to the number of men or women who belong to them. One of the types of households which are most severely burdened with expenditure on energy carriers are households run by a single person [30]. There are four times more energy poor households inhabited only by women (one or more) than households inhabited by only by men (one or more)—in the proportion of 481,500 to 121,200. Energy poor women, on the other hand, more often live-in households where there are no men (usually they are single-person households). A greater share of energy poor women, compared to the share of energy poor men within these groups, can be observed in the case of households: inhabited by one person (women: 11.3%, men: 3.2%); pensioners (women: 7.5%, men: 5.4%); retirees (women: 21.6%, men: 16.5%), single mothers with dependent children and other persons (women: 6.4%, men: 4.4%); people living in multi-family buildings (women: 45.5%, men: 41.3%); people living in apartments with an average area of 30–60 m^2 (women: 37.2%, men: 34%) [31]. An empirically significant negative causal relationship was found between the indicators of multidimensional energy poverty and health for women [32].

3.1. The Future Labor Market in Coal Sector

The decarbonization of the EU-27 is expected to result in the loss of approximately 76,000 jobs in coal mines and plants by 2025; the number is expected to double with more than 154,000 job losses projected by 2030 [14]. The withdrawal of coal will have different effects in different regions, it is making sure that no one is left behind, a 'just transition' for workers and communities as the world's economy responds to climate change was included as part of the 2015 Paris Agreement on climate change [14]. In the EU, coal infrastructure is present in 108 regions, and according to estimates the coal sector currently provides 237,000 jobs, mainly in coal mining—185,000 jobs. The largest number of people employed in the coal sector is found in the following European Union countries: Poland, Germany, the Czech Republic, Romania, Bulgaria, Greece, and Spain—this gives nearly 200,000 jobs directly related to coal, and it is in these regions that the largest job losses can be expected.

The transition to a decarbonized economy is a chance for the further development of green jobs, which refer to those with indirect or direct influence on environmental benefits [32]. There are forecasts which focus on green jobs according to their own definition, i.e., OECD, EU, UNEP, ILO. According to the ILO, changes in energy production can create 18 million jobs throughout the world economy. Employment in the EU-28's environmental economy rose from 2.8 million full-time equivalents (FTE) in 2000 to 4.5 million full-time equivalents in 2016 (ILO, 2018). Those include a shift towards renewable energy sources, as they have the potential to generate employment opportunities. Renewable energy production is more labor intensive than conventional energy production in delivering the same amount of energy output [33]. Employment in the renewable energy sector (wind, photovoltaic, biomass) can be divided into construction and operation. Directly and indirectly, there is about 1.51 million employees in renewable energy sector in the

European Union, what represents a gross growth of 67,000 jobs (+4.6%), of which Poland's increase is 11,900 jobs (16%) between 2017 and 2018 [31–38].

The role of hard coal in the Polish energy sector will gradually decrease. The transition to a low-carbon economy in the coming years will be a major challenge both for the country and for regions with a mining tradition. Poland will have to face not only the energy transformation, but also a significant transformation of the labor market. In 2018, Poland adopted its strategy for the coal sector in Poland until 2030, in 2019 Poland approved its National Energy and Climate Plan for the years 2021–2030 (NECP) which stipulated its objectives relating to increased energy efficiency and decarbonization, and in February 2021 it adopted Energy Policy 2040 in which a diminishing role of coal in energy generation and heating is foreseen, but in which also activities connected with Just Transition and a reduction in energy poverty are emphasized.

What should be taken into consideration during planning for the Just Transition is the multiplier effect, which is defined as the situation in which as the situation in which a rise of spending generates an increase in national income, which is bigger than initial amount spent. The employment multiplier effect defines the quantity of direct, indirect, and induced jobs created or lost in the area, which can be observed in Figure 3. The mining multiplier effect means that each loss of a mining job will cause several more workers to lose their jobs. At the end of December 2019 there were 83,297 people employed in hard coal mining in Poland, 64,259 of them working underground. A total of approximately 400,000 employees are employed in the mining facilities sector. In total, that provides nearly half a million jobs and, thus, it confirms the popular slogan that one job in mining generates three or four jobs in its surroundings. Job reduction in mining is associated with a reduction in jobs in sectors providing goods and services for mines. The number of employees associated with the activity undertaken by the mining sector is 56,700, i.e., about 0.4% of the number of employees in Poland.

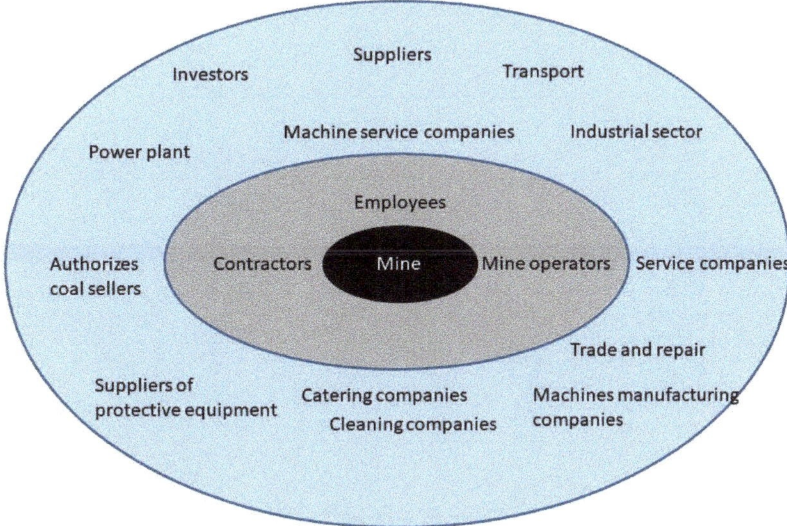

Figure 3. Jobs associated with mining divided into those in the close vicinity and those in the more distant surrounding area.

In accordance with the principles of Just Transition, the energy transformation process should be based on the following tools:

- Retirement;
- leaving mining;

- change of position inside the plant;
- relocation to another establishment;
- shifting to work on closing the mine;
- re-qualification for a profession other than mining.

In the terms of energy poverty, the potential for re-qualification for a job in the renewable energy sector should be a priority, but the requalification of women should also be taken into consideration. The relationship between national energy policy, which is divided into 6 transformation periods and employment rate was illustrated in Figure 4.

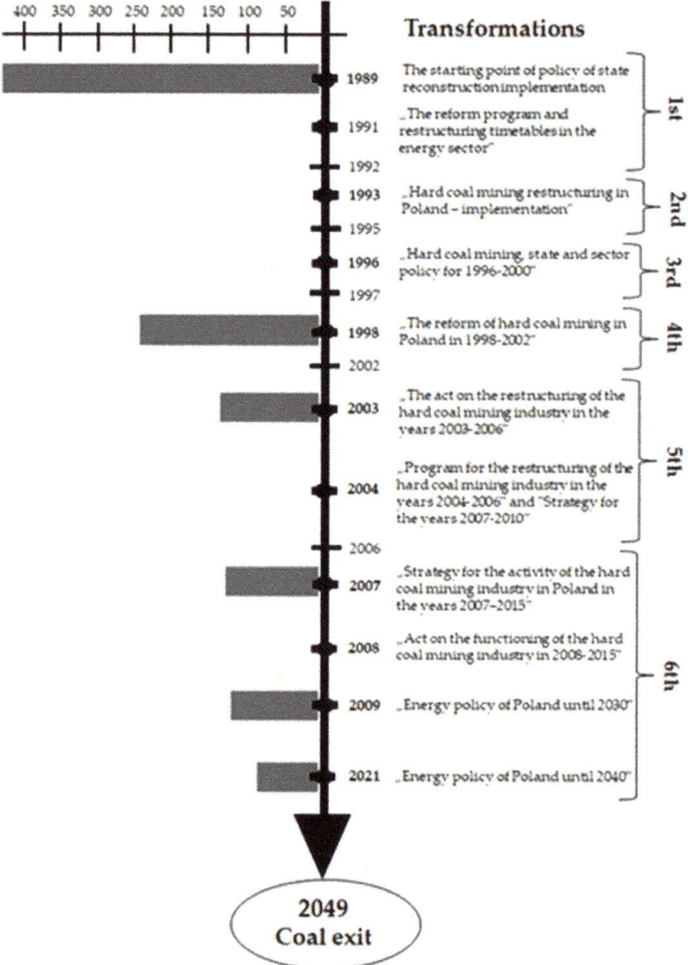

Figure 4. The relationship between national energy policy and employment rate.

3.2. Just Transition in the Silesian Voivodeship

The Silesian Voivodeship is in southern Poland. It is one of the smallest (2.8% of Poland) but the most densely (12% of Poland) populated voivodeship in Poland, where over 77% of region's population live in cities and towns. Moreover 77.6% of the population aged 25 to 54 years, and over 19% of the population is within the economically post-productive age. It is also the most industrial region in the country, almost 12.4% of the Polish GDP is being generated here what ranks the region on the second position within the country. Even though most of the Polish coal is mined in Silesia, the role of mining has been diminishing. However, 74,500 workers are employed in the hard coal mining industry mainly men, creates 7% of the vale added in the region. It is estimated that 4 time more, i.e., 280,000 employees are employed in the mining facilities sector, of which many women.

The process of closing mines in the Silesia Region planned will result in wide-ranging social and economic changes. The Silesia is currently at a turning point in its transformation, and 2021 is a year of key decisions for the future of the region and its inhabitants. The course of transformation will put pressure on the local economy, the labor market and social issues. Therefore, both the regional authority and the Polish government are working on plan of transformation, focusing mainly on possibility of creation jobs in different sectors, including renewable energy.

The declaration of the shutdown of the mining industry, concluded in an agreement between the government and the trade unions, begins a new stage in the comprehensive transformation process. The industries with a chance of replacing mining in the future are construction, industrial processing, and energy, they are part of the region's specialization, and the required skills, knowledge, and competences are part of the education profile of people employed in the mining industry. Other activities that may offer attractive jobs are logistics and remediation activities [39–42]. According to preliminary estimates, approximately 24–41,000 jobs may be created in the industries cited by 2030. Another 43–55,000 jobs may be created in activities diversifying the regional economy, such as development of business services. The creation of new jobs should be supported by strong incentives for investors who are ready to locate their activities in mining communes, support offered to existing and new entrepreneurs for the creation of high-quality jobs, support for the development of science and new education courses, and appropriate spatial policy.

According to the data, the number of unemployed people in Silesia who were registered with county labor offices in December 2020 totaled to 91,032 people and was 1545 people higher than in the previous month. Women predominate (51.4%) in the population of the unemployed in the Silesia Region (Statistics Research and Analysis Team). As of 31 December 2020, 49,851 unemployed women accounted for 54.8% of the total number registered (a month earlier, the corresponding share was similar and amounted to 54.9%).

Men and women often work in different professions and industries and have different average wages and working hours. The reasons for this fact lie in both historical and biological conditions. It was mainly men who were present on the Polish labor market as women gradually entered it, a process which took place after World War II and which was related to the dynamic development of the economy. The share of women in the number of employees in Poland has been steadily increasing since 1950 (Figure 5).

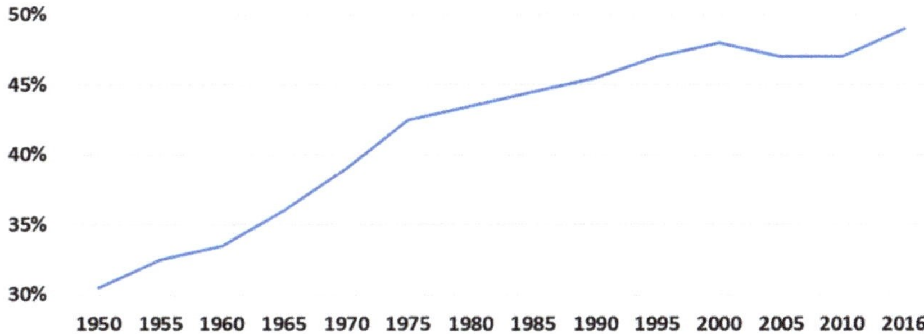

Figure 5. Share of women in the total number of employees in Poland, based on GUS, women and men in the labor market (2004–2018).

Sociological research conducted in the context of contemporary cultural changes, consisting in the weakening of the cultural pattern of pressure directed at women to see their role in life primarily in terms of being wives and mothers, indicates the barriers to the professional advancement of women that still exist, as well as the problems that result from combining social roles. This situation applies to women in Poland, especially women in the Silesia Region [43]. This is mainly due to the specific nature of the region. Traditionally, in the Silesian family, men played the role of breadwinner, and the miner's "honor" prevented his wife from working. An external factor influencing the role of women in Silesia was the almost exclusive availability of jobs in the mining industry, which permanently consolidated the cultural division of social roles. Additionally, one should consider that the qualifications of miners' wives were rather low—most of them had vocational education, a few had secondary education. When it comes to working women, before World War II, Silesian women mainly took up jobs in mines, steel mills, and factories. The zinc smelting industry employed the most women. In the 1940s and 1950s, women took up employment in industry, mostly performing auxiliary work. However, the living standards of Silesian families were mostly provided by male workers. In the 1960s, more and more girls graduated from vocational school, which triggered changes in the professional structure of the region. In the 1970s, the percentage of professionally active women increased, some sectors, such as trade, education, as well as health and social welfare were dominated by women. An important factor improving the education of women in Silesia was the establishment of the University of Silesia in 1968 which provided education in the humanities. The period of systemic transformation after 1989 subjected the Silesian family to a serious test. Phenomena such as loss of one's job, unemployment, and impoverishment affected Silesian families [15]. The employment structure with the division into gender between 2004 and 2018, can be seen in Figure 6.

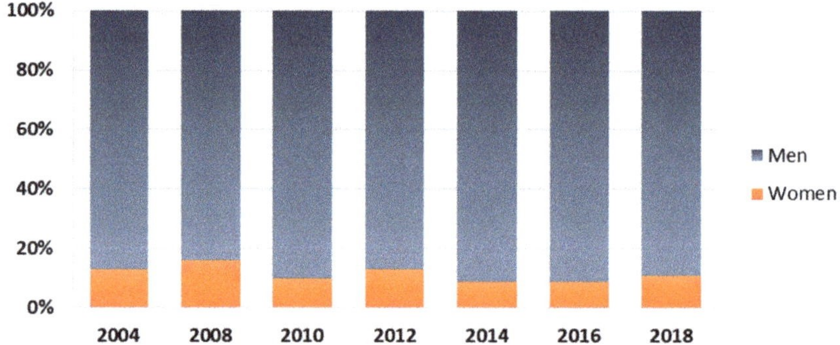

Figure 6. Mining and quarrying—employment with divisions into men and women, based on GUS, women, and men in the labor market (2004–2018).

In Poland, the ban on employing women underground in mines introduced in 1975 resulted from the Labor Code supplemented by executive regulations. At the time of Poland's accession to the European Union, the Polish labor law limited the scope for employing women underground in mines; it did not cover work performed in certain legal situations (permitted work). After Poland's accession to the European Union, the maintenance or lifting of this ban became an important problem, and its resolution depended on the outcome of proceedings pending before the European Court of Justice regarding Austrian regulations, which contain a prohibition like that in force in Poland.

3.3. Silesia of the Future

The process of restructuring Silesia should consider the fact that the core of Silesian identity is related to hard coal. The inhabitants of Silesia will also experience a profound cultural change, which affects the ethos of conscientious work, reliability, and love of family that is so important to the inhabitants of the region and which is the ethical code of the

Silesian population. It is important that the reform takes place with public acceptance of its goals and of the stages in its implementation. It must be remembered that the cost of the necessary social changes is one of the issues that worries the inhabitants of Upper Silesia the most. The process of closing the mining industry is inevitable, but in the name of fairness, solidarity, and sustainable development, this process should be more closely linked with the process of supporting mining municipalities so that cities built on coal and for the coal industry, as compensation for incurred costs and losses (including environmental), receive post-industrial infrastructure to be used for scientific, educational, social, cultural, and commercial purposes.

Replacing coal will accelerate the development of renewable energy, energy storage technologies, and other technologies that stabilize the operation of the power system. The key challenge of a just transition will be the equal distribution of costs and benefits resulting from technological change. For the Silesia Region, this should mean replacing jobs in declining industries with jobs in modern industry and services, especially in industries related to energy transformation. It is necessary to carefully discern not only the needs but also the possibilities of new labor markets, so that they are best suited to the emerging modern economy, but also to take into account the age structure and education of women.

4. Discussion

According to the SDG, more precisely to goal number 5—Gender equality according to which ending all discrimination against women and girls is the priority in terms of human rights. What should be strongly underlined is that, despite the fact that women are present at the labor market, still in some regions there are significant inequalities, with women systematically denied the same working rights as men. In the era of late modernity, developed capitalism, and the democratic postulates of participation, a special place in the process of just transformation of Silesia should be devoted to women. The role women play in the future labor market should be defined. As part of Just Transition, a serious discussion about women's expectations of employment opportunities in restructuring areas should be considered. It is necessary to carefully discern not only the needs but also the possibilities of new labor markets, so that they are best suited to the emerging modern economy, but also to take into account the age structure and education of women. Counteracting the professional exclusion of women in Silesia is extremely important in the context of energy poverty. It is true that the analysis of the gender distribution in the population and in the group of the energy poor shows that they are almost identical, yet, despite the lack of differences between men and women in the percentage of energy poverty at the general level, differences can be noted in the distribution of results for gender depending on individual categories.

For many reasons, for the global, national, and regional market, the restructuring of the coal industry, and, thus, the socio-economic transformations in the Silesia Region, are inevitable. They are part of the study with zero emission future of the European Union. It is worth adding that this is not the first time Silesia will undergo restructuring. In the years 1990–2000, the process of restructuring of the hard coal mining industry commenced, which, since 1993, was implemented on the basis of economic and social assumptions of restructuring programs approved by the Economic Committee of the Council of Ministers. The economic changes in Poland after 1989 in the hard coal mining industry began with the restructuring of the management of this sector. In 1990, the structures supervising the mines were liquidated, the Hard Coal Community, which was established in place of the previously liquidated Ministry of Mining and Energy, ceased to function. Hard coal mines have become independent state-owned enterprises. In the years 1989–2000, employment in the hard coal mining industry decreased by 2600 thousand employees. In the years 1989–1992, approximately 25,000 people left the mining industry annually. The high rate of decline in employment continued in the next two years (1993–1994). Employment decreased by 51,700 people, i.e., by 15.1%. In the following years, the pace of reducing employment was lower, in total 45,100 employees left their jobs. In the years 1998–1999, as a result of the application of the Mining Social Package instruments, there was a significant increase in the number of employees leaving jobs. The total decrease in employment in two years amounted to 697,000 people. In 2000, the decline in employment amounted to a further 18,600 employees. These changes caused several socio-economic problems, the consequences of which are felt in Silesia to this day.

Although Silesia is a region of Poland characterized by a low percentage of people suffering from energy poverty, it is also a region with a certain specificity. It is a region that is the largest mining area in Poland. In the perspective of energy poverty, which particularly affects single women living alone in Poland, it is extremely important in the transformation process to focus on women who will lose their jobs due to the changes or their working husbands will be subject to layoffs.

5. Conclusions

In this paper, we discuss energy transformation in Poland, and its influence on the energy poverty. As according to the data energy poverty in Poland mainly concerns households inhabited by single women, we are focusing on place and role of women in the future job market at Silesian Region. During working on the paper gap in research in this field was identified, what is especially important in the light of SDG 5.

An in-depth study on energy policy development pointed to the lack of sufficient social emphasis in the document strategy, as it is generally focused mostly on energy safety, transformation towards RES, and minimalization of the environmental impact. According to the authors there is a need to fulfil the mechanisms of Just Transition with a special agenda for the future place and role of women in new emerging job market. As it is single unemployed women who are in Poland the most vulnerable to energy poverty, we claim that a special hub for women as part of Just Transition should be created.

The number of women directly employed in mining in Poland is 7375, if we take under consideration the number of women who works in the nearest surrounding of mines this number will increase by another 5015 (Figure 7). Still if we take under consideration multiplier effect, quantity of direct, indirect and induced jobs created t in the area, we should multiply the total number of women employees in the mining sector by a factor of 4. Moreover, when it comes to future occupational restructuring, it should be underline that the lowest percentage of women employed in mining is those who works underground. Therefore, future restructuring programs should be designed for a wide professional scale; from female workers, through administrative, cooks, engineers, and cleaners, etc.

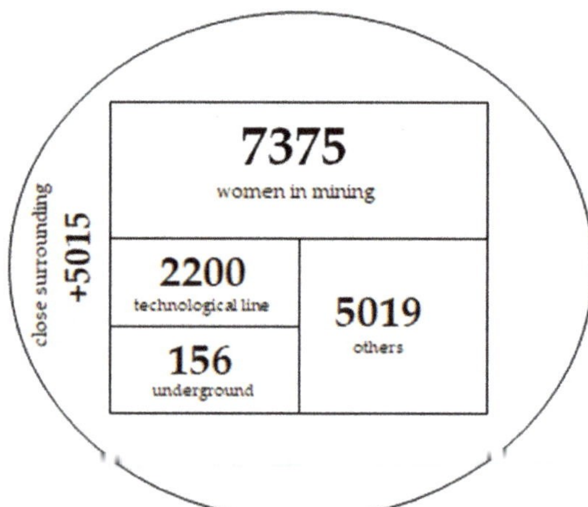

Figure 7. Number of women employed in mining, based on GUS, women, and men in the labor market (2004–2018).

The aim of the hub, which is a form of preventive response to the future problems of women in Silesia, would be: assistance in reskill from mining to different forms of employment, assistance in contact with reskilling course providers, psychological support do not give up approach, training in re-enter labor market (e.g., support in CV creation, best practices for the job interview, dress code, how to use social media), cooperation with the reskilling course providers, engaging issues that affect the living standards of Silesian woman's, promotion of sustainable development approach, support on job reorientation, provide technical support, assistance and programs to educators, industry, and local government.

Author Contributions: Conceptualization, J.K. and O.J.; methodology, J.K. and O.J.; formal analysis, J.K. and O.J.; investigation, O.J.; resources, O.J.; data curation, J.K. and O.J.; writing—original draft preparation, O.J.; writing—review and editing, J.K.; visualization, O.J. All authors have read and agreed to the published version of the manuscript.

Funding: This research was funded by the Res-skill Project (Reskilling coal industry workers for the renewable energy sector), co-funded by the Erasmus + Program of the European Union and under subvention funds for the Faculty of Management and by program "Excellence Initiative—Research University" for the AGH University of Science and Technology.

Institutional Review Board Statement: Not applicable.

Informed Consent Statement: Not applicable.

Data Availability Statement: Not applicable.

Conflicts of Interest: The authors declare no conflict of interest.

References

1. Boguszewski, R.; Herudziński, T. *Ubóstwo Energetyczne w Polsce*; Pracownia Badań Społecznych SGGW: Warszawa, Polska, 2018; pp. 23–25.
2. Owczarek, D. *(Nie)równy Ciężar Ubóstwa Energetycznego Wśród Kobiet i Mężczyzn w Polsce*; ClientEarth: Warszawa, Polska, 2016; pp. 5–7.
3. Boardman, B. *Fuel Poverty: From Cold Homes to Affordable Warmth*; Belhaven Press: London, UK, 1991; pp. 1071–1072.
4. Bouzarovski, S.; Petrova, S.; Sarlamanov, R. Energy poverty policies in the EU: A critical perspective. *Energy Policy* **2012**, *49*, 76–82. [CrossRef]
5. Buzar, S. *Energy Poverty in Eastern Europe: Hidden Geographies of Deprivation*; Routledge: Surrey, UK, 2007; p. 47.
6. Reddy, A.K.N. Energy and Social Issues. In *World Energy Assessment: Energy and the Challenge of Sustainability*; United Nations Development Programme: New York, NY, USA, 2000; pp. 179–185.
7. Halkos, G.; Gkampoura, E.-C. Evaluating the effect of economic crisis on energy poverty in Europe. *Renew. Sust. Energ. Rev.* **2021**, *144*, 11. [CrossRef]
8. Bardazzi, R.; Bortolotii, L.; Pazienza, M.G. To eat and not to heat? Energy poverty and income inequality in Italian regions. *Soc. Sci.* **2021**, *73*. [CrossRef]
9. Selecting Indicators to Measure Energy Poverty. Available online: https://ec.europa.eu/energy/sites/ener/files/documents/Selecting%20Indicators%20to%20Measure%20Energy%20Poverty.pdf (accessed on 2 September 2020).
10. Policy Report, Energy Poverty and Vulnerable Consumers in the Energy Sector across the EU: Analysis of Policies and Measures. 2015. Available online: https://ec.europa.eu/energy/sites/ener/files/documents/INSIGHT_E_Energy%20Poverty%20-%20Main%20Report_FINAL.pdf (accessed on 10 July 2020).
11. Bouzarovski, S. *Energy Poverty. (Dis)Assembling Europe's Infrastructural Divide*, 1st ed.; Springer Nature: London, UK, 2018.
12. Addressing Energy Poverty in the European Union: European State of Play and Action. Available online: https://www.energypoverty.eu/sites/default/files/downloads/publications/18-08/paneureport2018_final_v3.pdf (accessed on 10 January 2021).
13. The Health Impacts of Cold Homes and Fuel Poverty. Available online: http://www.instituteofhealthequity.org/resources-reports/the-health-impacts-of-cold-homes-and-fuel-poverty/the-health-impacts-of-cold-homes-and-fuel-poverty.pdf (accessed on 10 July 2020).
14. Paris Agreement, United Nations. Available online: https://unfccc.int/sites/default/files/english_paris_agreement.pdf (accessed on 1 February 2021).
15. Swadźba, U. Śląskie wartości-praca, rodzina i religia: Geneza, trwanie i zmiany. *Górnośl. Stud. Socjol.* **2014**, *5*, 22–41.
16. GUS. Kobiety i Mężczyźni na Rynku Pracy. 2018. Available online: http://eregion.wzp.pl/sites/default/files/kobiety_i_mezczyzni_na_rynku_pracy_2018.pdf (accessed on 10 July 2020).

17. Report from the Commission to the European Parliament, the Council, the European Economic and Social Committee and the Committee of the Regions. Available online: https://ec.europa.eu/transport/sites/transport/files/legislation/com20190120.pdf (accessed on 10 July 2020).
18. Directive 2009/72/ECof the European Parliament and of the Council of 13 July 2009 Concerning Common Rules for the Internal Market in Electricity and Repealing Directive 2003/54/EC. Available online: https://eur-lex.europa.eu/legal-content/EN/TXT/PDF/?uri=CELEX:32009L0072 (accessed on 10 July 2020).
19. Directive 2009/73/EC of the European Parliament and of the Council of 13 July 2009 Concerning Common Rules for the Internal Market in Natural Gas and Repealing Directive 2003/55/EC. Available online: https://eur-lex.europa.eu/legal-content/EN/ALL/?uri=CELEX%3A32009L0073 (accessed on 10 July 2020).
20. Third Energy Package. 2009. Available online: https://ec.europa.eu/energy/en/topics/markets-and-consumers/market-legislation (accessed on 11 November 2019).
21. European Economic and Social Committee (EESC). Opinion of the European Economic and Social Committee 'for Coordinated European Measures to Prevent and Combat Energy Poverty'. *Off. J. Eur. Union* **2013**, *56*, 1–8.
22. Energy Policy of Poland until 2040 (EPP2040). Available online: https://www.bing.com/search?q=Energy+Policy+2040+&form=ANNTH1&refig=c3e3d56b41934906b738fbce430e7e56 (accessed on 10 March 2021).
23. Energy Policy of Poland until 2030. Available online: https://www.cire.pl/item,38932,2,0,0,0,0,0,polityka-energetyczna-polski-do-2030-roku.html (accessed on 10 April 2021).
24. Energy Policy of Poland until 2025. Available online: https://www.cire.pl/item,15970,2,0,0,0,0,0,polityka-energetyczna-polski-do-2025-roku.html (accessed on 10 April 2021).
25. Energy Policy of Poland until 2020. Available online: https://www.cire.pl/item,800,2,0,0,0,0,0,zalozenia-polityki-energetycznej-polski-do-2020-roku---skorygowane-zalozenia-przyjete-przez-rade-ministrow-2-kwietnia-2002r.html (accessed on 10 April 2021).
26. Clean Energy for all Europeans. Available online: https://ec.europa.eu/energy/topics/energy-strategy/clean-energy-all-europeans_en (accessed on 10 March 2021).
27. Sen, A. *Development as Freedom*; Oxford University Press: Oxford, UK, 1999; pp. 22–23.
28. Brock, G. *Global Justice a Cosmopolitan Account*; Oxford University Press: New York, NY, USA, 2009; pp. 18–20.
29. Archibugi, D.; Held, D. *Cosmopolitan Democracy: Paths and Agents. Ethics International Affairs*; Cambridge University Press: Cambridge, UK, 2011; Volume 25, pp. 433–461.
30. Piekut, M. Energy consumption in polish one person households. *Energies* **2020**, *13*, 5699. [CrossRef]
31. Kiewra, D.; Szpor, A.; Witajewski-Baltvilks, J. *Sprawiedliwa Transformacja Węglowa w Regionie Śląskim. Implikacje dla Rynku Pracy*; IBS Research Reports 2019; Instytut Badań Strukturalnych: Warszawa, Polska, 2019; pp. 32–39.
32. Structural Change in Coal Phase-Out Regions. Available online: https://www.espon.eu/sites/default/files/attachments/Policy%20Brief%20structural%20change%20in%20coal%20phase-out%20regions.pdf (accessed on 10 March 2021).
33. Uchwała w Sprawie "Polityki Energetycznej Polski do 2040 r". Available online: https://www.gov.pl/web/polski-atom/uchwala-w-sprawie-polityki-energetycznej-polski-do-2040-r (accessed on 10 March 2021).
34. Aceleanu, M.I.; Pociovalisteanu, D.M.; Serban, A.C.; Grecu, E. The link between environment and employment for a sustainable development. In Proceedings of the International Scientific Symposium, Information Society and Sustainable Development, Azuga, Romania, 24–25 April 2015.
35. Babonea, A.M.; Joia, R.M. Transition to a green economy—A challenge and a solution for the world economy in multiple crisis context. *Theor. Appl. Econ. J.* **2012**, *10*, 90–101.
36. United Nations Environment Programme (UNEP). Towards a Green Economy: Pathways to Sustainable Development and Poverty Eradication. Available online: http://sustainabledevelopment.un.org/index.php?page=view&type=400&nr=126&menu=35 (accessed on 17 January 2015).
37. International Labour Organization (ILO). What is a Green Jobs? Available online: http://www.ilo.org/global/topics/green-jobs/news/WCMS_220248/lang--en/index.htm (accessed on 14 December 2014).
38. United Nations Environment Programme; International Labour Organization; International Organisation of Employers; International Trade Union Confederation. Green Jobs: Towards Decent Work in a Sustainable, Low-Carbon World. 2008. Available online: https://www.greengrowthknowledge.org/sites/default/files/downloads/resource/Green_Jobs_Towards_Decent_Work_in_a_Sustainable%2C_Low-Carbon_World_Worldwatch_Institute_0.pdf (accessed on 29 April 2021).
39. Low-Carbon World. Available online: http://www.unep.org/PDF/UNEPGreenJobs_report08.pdf (accessed on 14 December 2014).
40. Thomson, H.; Snell, C.; Liddell, C. Fuel poverty in the European Union: A concept in need of definition? *People Place Policy* **2016**, *10*, 5–24. [CrossRef]
41. Karpinska, L.; Śmiech, S. Breaking the cycle of energy poverty. Will Poland make it? *Energy Econ.* **2021**, *94*. [CrossRef]
42. Alves Dias, P.; Kanellopoulos, K.; Medarac, H.; Kapetaki, Z. *EU Coal Regions: Opportunities and Challenges Ahead*; Publications Office of the European Union: Luxembourg, 2018. [CrossRef]
43. Sulich, A.; Zema, T. Green jobs, a new measure of public management and sustainable development. *Eur. J. Environ. Sci.* **2018**. [CrossRef]

Article

Socioeconomic Factors Influencing the Prosumer's Investment Decision on Solar Power

Patrick Rausch and Michał Suchanek *

Faculty of Economics, University of Gdańsk, Armii Krajowej 119/121, 81-824 Sopot, Poland;
patrickrausch@hotmail.com
* Correspondence: michal.suchanek@ug.edu.pl

Abstract: This paper identifies socioeconomic factors that are supposed to impact the investment decision of private households in Germany regarding small-scale solar PV (photovoltaic) systems. In 2022, the last nuclear power plant will phase out and the end of coal-fired power plants is fixed for 2038. Thus, the legislator is mandated to foster the addition of renewable energy capacities to close the gap fossil fuels and nuclear power leaves. Some share of the renewable energies could be from private households that mainly invest in small-scale solar PV systems. To stimulate investments, it is necessary to identify factors that are important for the investment decisions of private households. Within this paper, secondary socioeconomic data for the period from 2009–2018 was compiled. In order to identify the latent variables, a factor analysis was conducted. The results state five factors that are supposed to impact the investment decisions of the prosumer in Germany.

Keywords: renewable energy; prosumer decision; factor analysis; investment decision

check for
updates

Citation: Rausch, P.; Suchanek, M. Socioeconomic Factors Influencing the Prosumer's Investment Decision on Solar Power. *Energies* **2021**, *14*, 7154. https://doi.org/10.3390/en14217154

Academic Editor: David Borge-Diez

Received: 28 September 2021
Accepted: 26 October 2021
Published: 1 November 2021

Publisher's Note: MDPI stays neutral with regard to jurisdictional claims in published maps and institutional affiliations.

1. Introduction

Recently, climate change has become a topic that is top of the agenda of legislators and civil society worldwide. At the world climate conference in Paris, 2015, most of the governments confirmed these severe problems. Today, 189 countries have ratified the Paris Agreement.

In Germany, the energy transition is in progress, meaning the expiration of nuclear power plants in the year 2022 [1]. Furthermore, in 2020, the German legislator decided to phase out coal-fired power plants by 2038 [2]. The gap that fossil fuels and nuclear power leave is supposed to be closed by renewable energies, especially wind and solar power [3]. In order to foster the renewable energy systems (RES), the renewable energy law, Erneuerbare-Energien-Gesetz (EEG) was introduced 2000 in Germany to subsidize and scale-up RES [4]. As a consequence, costs for the energy turnaround increased and the political sector was encouraged to step-down the guaranteed feed in tariffs for installations [3]. Due to the lowered subsidies and the intention of the legislator to integrate the RES, thus, also solar PV systems, into the regular electricity market system, investments decreased significantly [5]. From 2012 onward, the yearly added capacities of solar PV systems started to decrease, from 7.600 MWp (megawatt peak) in 2012 to 1.500 MWp in 2016. Since 2019, investments started to recover and reached almost 4.000 MWp [6]. Nevertheless, the current investments in solar PV systems are not enough for Germany's energy transition.

In order to increase the yearly added capacities and reach the proclaimed targets of Germany's energy transition, investments of private households could help to close the current gap. Data from December 2019 from the German Federal Network Agency reveal that approximately 34% of the investments in December 2019 falls upon small-scale solar PVs (<10 KWp) [7]. Solar plants < 10 KWp are mostly owned by private households in Germany as the legislator implemented special subsidies for private households investing in solar plants < 10 KWp [8]. In contrast to investments from companies, funds, etc., investments from private households have been quite stable over the last decade, thus, the prosumer is a stable source of newly added capacities in the field of solar PV systems.

The focal point of this paper lies in socioeconomic factors that are supposed to influence the investment decisions of the prosumer. In this regard, we considered the period from 2009–2018 for 20 independent variables. The selection of the variables was based on the transaction cost theory (TCT). To reduce the number of variables and eliminate collinearity, a factor analysis was conducted.

This paper could help to close research gaps regarding factors that impact the investment decision of private households and the implied decisions of legislators in Germany and worldwide. Firstly, as the identified factors are of a macroeconomic nature, decision makers in Germany would be able to align their policies in the field of renewable energy sources (RES) in a more precise manner. As the different states in Germany show different levels of socioeconomic factors, decision makers are supposed to align their legislation towards different regions in Germany. Secondly, the results could be applied not only in Germany, but worldwide, and help decision makers to foster the investments of private households, thus, make their inhabitants part of the energy transition.

There are vast numbers of authors and papers that are researching factors that influence the development or the perception of RES. Most of the papers are of a descriptive nature and describe a special analysis in some countries. For example, the USA, Finland and Germany [9–11]. There are just a few papers that have applied econometric models.

Bilgen, Keles and Kaygusuz (2008) conducted a case study on Turkish energy related environmental policies [12]. Ince, Vredenburg and Liu (2016) performed a panel analysis of renewable energy development in the Caribbean. Based on interviews and case studies, seventy-five factors were identified for a regression analysis. With the help of the quantitative analysis, five factors were identified as critical regarding the development of renewable energies [13]. Dui'c, da Graça Carvalho (2004) used a H2RES model on the example of an isolated island in the Madeira archipelago [14]. By using this model, they were looking for the optimization of the integration of hydrogen usage with intermittent renewable energy sources. Shahrestani, Yao, Luo, Turkeyler and Davies (2015) chose a different scientific approach to investigate the urban microclimate and their effects on urban renewable energy technology implementation [15]. An experimental design was used, measuring in a field study. This is similar to Bajic and Divić (2010), who used results of a field study measuring wind on the Adriatic coast to evaluate how wind energy technologies make sense in Croatia [16].

Some authors have applied a factor analysis in the field of RES recently. Within their studies, authors have e.g., researched factors that affect the willingness of the public to adopt RES and factors that influence innovations in the field of RES.

Makki and Mosley (2020), for example, researched the factors that affect the willingness of the public to adopt new RES technologies. According to the authors based on a literature review, 19 factors were supposed to impact the public willingness in regard to renewable energies. Afterward, an online survey was applied to collect random cross-sectional data of 416 participants by using the extracted factors. Subsequently, the key components were extracted by an exploratory factor analysis. In the end, five main categories were revealed clustering the nineteen extracted factors [17]. Another paper by He, Xu, Li and Zhao (2018), depicted factors that influence technological innovation in the field of RES [18]. Olanrewaju, Olubusoje, Adenikinju and Akintande (2019) were interested in the underlying factors of consumption of renewable energies and have conducted a panel model to dig out the relevant factors in Africa. To do so, annual data from 1990–2015 of five African countries was applied [19].

There are at least two recent studies that have researched macroeconomic factors impacting the development of RES.

Grzeszczyk, Izdebski, Izdebski and Wasinski (2021) researched the influence of socioeconomic factors on the development of RES in Poland [20]. Twenty factors, according to a factor tree analysis, were researched. The results suggests that EU as well as Polish legislation has a decisive impact on the development of RES in Poland where the EU and Polish legislation is coined by lobbying groups that either try to foster or block the extension of RES. Fatima, Li, Ahmad, Jabeen and Li (2021) tried to identify crucial factors that influence the development of RES. To do so, a content analysis as well as a modelling analysis was conducted, applying data collected via a questionnaire survey. The results state that a lack of good governance, renewable energy adaption as well as the governmental energy policies are relevant factors for the development of RES [21].

As the aforementioned studies prove, socioeconomic factors have an impact on the development of RES. According to Grzeszczyk, Izdebski, Izdebski and Wasinski (2021), the legislator is the main driver for investments within the market of RES [20]. This result is affirmed by the study of Fatima, Li, Ahmad, Jabeen and Li (2021) [21].

This paper is complementary to the stated papers, it is interested in socioeconomic factors that the legislator is able to adjust. Thus, we go one step further by assuming that the legislator has a decisive influence and the need to search for elevating screws to stimulate investments. In this context, the focal point lies on the prosumer. It is valuable to get deeper insight into the investments of private households as their demand regarding legislation is assumed to be different than from companies investing in the field of RES.

2. Materials and Methods

To select the variables that impact the investment behavior of the prosumer, TCT is applied. The main hypothesis of TCT by Williamson is the discriminating alignment hypothesis, where the governance structure serves as the dependent variable and the transactions the independent variable. This hypothesis has to be aligned in order to minimize costs [22]. We applied and modified TCT. Within this paper, transaction costs and governance structure serve as independent variables that are supposed to influence the investments of the prosumer. As seen in the study of Fatima, Li, Ahmad, Jabeen and Li, at the least, a lack of good governance has an impact on the development of RES.

In order to operationalize transaction costs and governance structure, indicators were applied to indicate either the level of the transaction costs or governance structure. According to a paper of Rindfleisch and Heide (1998) indicators are often applied to measure transaction costs and governance structure when secondary data is applied [23].

Starting with the governance structure, we assumed that three different groups of indicators are relevant when it comes to the display of the level of governance structure: demographic, social and political. The approach of different categories displaying the level of governance structure and transaction costs is also pursued by Bussolo and Whalley [24]. The authors have identified three major categories of transaction costs that are applied in the papers about political TCT.

The three subgroups are shortly explained before stating the 20 indicators.

Within this paper, the first group, demographic, entails factors such as sex, age, birth-death rates. Various authors have researched the influence of demographic factors on investment decisions and have confirmed that demographic factors matter [25]. We assumed that indicators such as BDR (Birth–Death Ratio) or EoL (Expectation of Life) were appropriate to display if a country or region displayed a positive or negative demographic trend. As the upfront investments and the payback period for solar PV systems is quite high for the prosumer, the assumption is that these indicators impact the development of RES.

The next group comprises social and political indicators. Different indicators such as the crime index and happiness index are subsumed under this group. The overall assumption is that, for example, the crime rate has an influence on the investment decision of individuals. In this context, Shanmugham and Ramya (2012) have shown that social factors have an influence on the investment decisions of individuals [26]. In addition, Ciobanu and Bahna (2015) have shown in their paper that social and political factors such as health expenditure and life expectancy matter regarding the investment decisions of individuals [27]. We assumed that investments with a long payback from the prosumer are completed in a safe and stable environment where inhabitants can save deposits and assume a social and political environment that allows investments with a long payback period.

The last group consists of the economic indicators. Because relevant upfront investments are needed to set up photovoltaic systems, factors such as household income are supposed to have an impact on the investment decisions of private households. In this regard, Geetha and Ramesh (2012) confirm the link between household income and investment decisions [25].

In the context of indicators that display the level of transaction costs, three indicators are chosen that fulfill the criteria of (a) being related to the field of renewable energies in a broader sense and (b) the data is available for the researched period.

Table 1 displays an overview about the applied indicators.

Table 1. Indicators.

Indicator	Source	Website
Demographic		
Birth–Death Ratio (BDR)	The data for this indicator is retrieved from Deutschland in Zahlen (Germany in numbers).	https://www.deutschlandinzahlen.de/tab/bundeslaender/demografie/natuerliche-bevoelkerungsbewegungen/geburtensaldo (accessed on 28 September 2021)
Expectation of Life (EoL)	The data is downloaded from a Statista pro account which could be used in the library of the University of Hamburg.	https://de.statista.com/statistik/daten/studie/820320/umfrage/lebenserwartung-in-bundeslaendern-nach-geschlecht/ (accessed on 28 September 2021)
Share of inhabitants living in cities > 100.000 in %	The data for this indicator has been retrieved in the following approach: Firstly, research on the websites of the particular German states has displayed which cities in the states entail > 100.000. In the next step, the population trend for these cities was researched on the websites of the cities for the relevant period. Afterward, within an excel sheet, the share of the people living in cities was calculated summing up all the people living in cities > 100.000 in the particular year divided by the entire population of the particular state.	The data is retrieved from the particular websites of the 16 states in Germany as well as homepages of the cities with more than 100.000 in order to retrieve the data for each year within the considered period.
Social and political		
Crime (Registered offences in Germany)	In this case, the primary data is issued by the Bundeskriminalamt (German Federal Office of Criminal Investigation).	https://de.statista.com/statistik/daten/studie/3459/umfrage/bundeslaender-nach-haeufigkeitszahl-von-straftaten-seit-2007/ (accessed on 28 September 2021)
Doctor density (number of inhabitants per doctor)	The data is extracted from Statista a report of the Bundesärztekammer (German Medical Association). The Bundesärztekammer is the head organization of the medical self-management in Germany.	https://www.bundesaerztekammer.de/fileadmin/user_upload/downloads/pdf-Ordner/Statistik2019/Stat19AbbTab.pdf (accessed on 28 September 2021)
Happiness Index	Collected from the "Glücksatlas" from a particular year. The Glücksatlas is issued by Bernd Raffelhüschen (Professor for financial economics) and Reinhard Schlinkert (director of dimap, a market research bureau). The Glücksatlas depicts the perceived happiness of inhabitants of all 16 German states.	https://www.dpdhl.com/de/presse/specials/gluecksatlas.html (accessed on 28 September 2021)
Saving deposits per capita in €	The data is retrieved from Statista. The primary source is the Bundesbank (the German central bank).	https://de.statista.com/statistik/daten/studie/203152/umfrage/spareinlagen-pro-kopf-nach-bundeslaendern/ (accessed on 28 September 2021)

Table 1. *Cont.*

Indicator	Source	Website
Spending per pupil	The data is retrieved from Deutschland in Zahlen. The primary source is the Statistisches Bundesamt (Federal Statistical Office).	https://www.deutschlandinzahlen.de/tab/bundeslaender/bildung/bildungsausgaben/staatliche-ausgaben-je-schueler (accessed on 28 September 2021)
Votes for far-right parties	The election results for the German states are retrieved from their websites.	The websites of all 16 German states are relevant.
Economic		
Constructed new flats per capita	Own calculation based on the constructed new flats per state and the population of the particular state. Both retrieved from Deutschland in Zahlen. The primary data is from Statistisches Bundesamt (Federal Statistical Office).	https://www.deutschlandinzahlen.de/tab/bundeslaender/infrastruktur/gebaeude-und-wohnen/neu-fertiggestellte-wohnungen (accessed on 28 September 2021)
Debt per inhabitant	The data is retrieved from Deutschland in Zahlen. The primary source is Statistisches Bundesamt (Federal Statistical Office)	https://www.deutschlandinzahlen.de/tab/bundeslaender/oeffentliche-haushalte/schulden/schulden-je-einwohner (accessed on 28 September 2021)
Filing of a patent per 100.000	The data is retrieved from Deutschland in Zahlen. The primary source is Statistisches Bundesamt (Federal Statistical Office)	https://www.deutschlandinzahlen.de/tab/bundeslaender/wissenschaft-forschung/patente/patentanmeldungen-je-100000-einwohner (accessed on 28 September 2021)
GDP per Capita	The data is retrieved from Deutschland in Zahlen. The primary source is Statistisches Bundesamt (Federal Statistical Office)	https://www.deutschlandinzahlen.de/tab/bundeslaender/volkswirtschaft0/bruttoinlandsprodukt/bruttoinlandsprodukt-je-einwohner (accessed on 28 September 2021)
Household income per inhabitant	The data is retrieved from Deutschland in Zahlen. The primary source is Statistisches Bundesamt (Federal Statistical Office)	https://www.deutschlandinzahlen.de/tab/bundeslaender/finanzen/einkommen-verdienste/haushaltseinkommen-je-einwohner (accessed on 28 September 2021)
Industry's share of the total employees in %	Own calculation based on the employees in the industry per state and the overall employed people of the particular state. Both retrieved from Deutschland in Zahlen. The primary source is Statistisches Bundesamt (Federal Statistical Office)	https://www.deutschlandinzahlen.de/tab/bundeslaender/branchen-unternehmen/industrie/beschaeftigte-in-der-industrie (accessed on 28 September 2021)
Share of homeowners in %	Statista	
Indicators displaying the level of transaction costs		
Remuneration per employee in the construction trade p.a. in 1000€ in Germany	The data is retrieved from Deutschland in Zahlen. The primary source is Statistisches Bundesamt (Federal Statistical Office)	https://www.deutschlandinzahlen.de/tab/bundeslaender/branchen-unternehmen/bau/entgelte-im-bauhauptgewerbe (accessed on 28 September 2021)
Research expenditures for renewable energies in relation to GDP in €/Mio €	The data is retrieved from the website föderal erneuerbar (federal renewable) which is administered by the Agentur für erneuerbare Energien (Agency for renewable energies), a German think tank.	https://www.foederal-erneuerbar.de/landesinfo/bundesland/D/kategorie/forschung/auswahl/228-forschungsausgaben_d/#goto_228 (accessed on 28 September 2021)
Spending for governmental employees per capita	The data is retrieved from Deutschland in Zahlen. The primary source is Statistisches Bundesamt (Federal Statistical Office)	https://www.deutschlandinzahlen.de/tab/bundeslaender/oeffentliche-haushalte/einnahmen-und-ausgaben-des-staates/personalausgaben-je-einwohner (accessed on 28 September 2021)

These variables, while being potentially significant descriptors of customer behavior regarding the photovoltaic market, can also be internally correlated. Therefore, our aim was to aggregate those observable variables, resulting from the content analysis based on the ideas of transaction theory into new, latent variables, presenting not the individual occurrences, but a more complex set of internally independent phenomena. This led to a decision to aggregate the observable variables into latent variables with the use of factor analysis. As this paper researches the 16 German states, indicators have been applied where the data is available for the 16 states during the researched period.

For most of the factors, the websites deutschlandinzahlen.de and statista.de (premium account) were used. These websites offer aggregated secondary data and use the German Federal Statistical Office as their primary source.

An exploratory factor analysis (EFA) approach was used in order to group the aforementioned variables which had been identified for the purpose of the study. EFA is a statistical technique of dimension reduction used to minimize a larger number of variables into a set of internally correlated factors. The new factors represent the underlying latent characteristics present in a wider set of variables, thus, compiling sets of variables which can be partially correlated into aggregated factors representing a set of distinct characteristics [28]. EFA was initially developed as a dimension reduction technique used to prepare an application of methods such as regression while minimizing the risk of collinearity. However, it is also widely used as an exploratory technique used to analyze the internal variability of the dataset. The first step of the EFA is usually to verify the Kaiser–Meyer–Olkin (KMO) criterion and to run the Bartlett's test of sphericity to verify the possibility of applying EFA to the given dataset. If the test is significant and the eigenvalues which are the basis for the criterion are higher than 1 then the extracted factors can be taken into analysis. In this case the classic principal components analysis (PCA) method was used to extract the factors [29]. The extracted factors were analyzed based on the eigenvalues and on the total variance explained and then the internal structure of the factors was analyzed, looking at the factor loadings for each item within the factors. There is a lot of discussion regarding the minimal acceptable value of the factor loadings for the item to be considered significant [30]. We accepted a factor loading for an item higher than 0.5 as a relatively good representation of a given item within the factor. Ultimately, 20 variables were taken into account with a total of 153 observations for individual German Lands collected for the years from 2009–2017. Table 2 illustrates the descriptive statistics for the variables which include mean, standard deviation as well as minimal and maximal values.

Table 2. Descriptive statistics for the 23 variables.

Variable	Mean	Std. dev.	Min	Max
Birth–Death	−20,587	40,664	−211,756	7036
Constructed new flats per capita	0.00253	0.00096	0.00066	0.00502
Remuneration per employee in the construction trade p.a.	27,937	3696	21,199	37,770
Crime	8348	2748	4868	16,414
Debt per inhabitant in €	9037	6480	454	32,405
Doctor Density (Inhabitants per doctor)	224	31	139	276
Expectation of life in years (Boys when born)	78	1	76	80
Filing of a patent per 100.000	39	34	7	144
GDP per capita in €	33,958	9248	20,482	64,567
Happiness index	6.98	0.21	6.52	7.43
Household income per inhabitant	19,792	2098	15,845	24,421
Share of homeowner in %	43.82	11.93	14.20	63.70
Research expenditures for renewable energies in relation to the GDP in €/Mio €	32.07	30.83	0.00	138.00
Savings deposits per capita in €	19,560	7766	10,140	40,770
Industry's share of the total employees in %	12.86%	3.70%	5.37%	19.93%
Share of inhabitants living in cities >100.000 in %	35.53%	31.15%	10.20%	100.00%
Spending per pupil in €	6543	850	4900	8900
Spending per student in €	7199	1093	5081	9727
Votes for far-right parties	4.89%	5.53%	0.00%	26.20%

EFA was used to reduce the identified variables into specific aggregated factors representing characteristics important in relation to the energy market. PCA was applied in order to identify the factors. Ultimately, as shown in Table 3, 20 items were aggregated into 5 factors which explain a total of 70.71% of initial variability, meaning that the loss from the variability in the initial data is relatively insignificant and acceptable [31].

Table 3. Eigenvalues for the extracted factors.

Factors	Eigenvalue	% of Variance Explained
Factor 1	5.49	27.45%
Factor 2	4.33	21.63%
Factor 3	1.59	7.95%
Factor 4	1.67	8.36%
Factor 5	1.07	5.34%

This indicates that the process of aggregation of the initial 20 items into the 5 new variables—factors—which are now not correlated with each other, has been successful and the new factors represent the wide array of determinants of customer behavior on the photovoltaic market. The internal structure of the factors and the character of their representation and meaning is presented in the following section.

3. Results

The five identified factors represent the main groups of variables affecting the energy market in Germany. Table 4 presents the factor loadings calculated for all the items within the five factors after the application of the PCA extraction method and the varimax rotation. Only factor loadings for items significantly represented in the factor are presented in the table to increase readability. These five factors together form the internal variety of the determinants of customer behavior regarding the photovoltaic market.

Table 4. Factor loadings for the extracted factors (varimax rotation).

Variable	Factor 1	Factor 2	Factor 3	Factor 4	Factor 5
Birth–Death					0.77
Constructed new flats per capita	0.67				
Remuneration per employee in the construction trade p.a.		0.88			
Crime	0.84				
Debt per inhabitant in €		0.79			
Doctor Density (Inhabitants per doctor)					
Expectation of life in years (Boys when born)	0.86				
Filing of a patent per 100.000	0.59			0.58	
GDP per capita in €	0.81				
Happiness index	0.80				
Household income per inhabitant	0.97				
Share of homeowner in %					
Research expenditures for renewable energies in relation to the GDP in €/Mio €					
Savings deposits per capita in €	0.61				
Industry's share of the total employees in %		−0.54		0.65	
Share of inhabitants living in cities >100.000 in %		0.95			
Spending per pupil in €			0.70		
Spending per student in €			0.68		
Spending per government employee in €		0.80			
Votes for far-right parties				−0.68	

Specifically, five factors which present the variability of these determinants from five different perspectives have been aggregated from the initial list of 20 items.

The first factor represents 27.45% of the initial variability and the items which are grouped together within it include: constructed new flats per capita, crime rate, life expectancy, patents filed, GDP per capita, happiness index, household income per inhabitant and the savings deposit. Given the general characteristics of the items presented in the factor it can be said to describe the overall socioeconomic level of a given region.

The second factor represents 21.63% of the total variance and groups together factors such as: remuneration per employee in the construction trade, debt per inhabitant, the industry's share of the total employees, spending per government employee and share of inhabitants living in cities above 100,000 citizens. This would indicate that this factor represents the urbanization level of a given region,

The third factor, which bulks together nearly 8% of the total variance demonstrates the education effort of a given region, grouping just two items: spending per pupil and spending per student, without any other items included.

Factor four groups together two very characteristic features of the industrialization of the region: the share of employees working in the industry sector as well as the number of patents filed per capita. The innovation represented by the patents is traditionally correlated with industry-heavy regions. Interestingly enough, within this factor, the share of citizens voting for far-right parties is also represented, with an opposite sign, representing an inverse correlation. This is a characteristic of Germany itself, where the far-right parties are traditionally more supported in the more rural and agricultural regions.

Lastly, the fifth factor, standing for slightly above 5% of the total variance consist of of one significant item, which is the birth–death variable, demonstrating that it seems to be heavily independent of all the other variables and warranting a variability of its own.

4. Conclusions

Based on the factor analysis, it could be stated that there are socioeconomic factors that impact the investment decisions of the prosumer in Germany. The first, second and fourth factors especially, could be interesting to analyze considering this paper. The findings of Grzeszczyk, Izdebski, Izdebski and Wasinski (2021), and Fatima, Li, Ahmad, Jabeen and Li. support this result.

The first factor could be seen as the quality of life. A lower level of the general quality of life is supposed to have an impact on the investment decision of private households. Looking on the underlying indicators reveal that these ones are difficult to adjust for the legislator in the short term. Thus, the legislator in Germany, and, of course, in other countries should intend to stimulate investments from regions and private households with a lower level of quality of life to distribute the subsidies in a more equitable way. Otherwise, the subsidies paid by the state are cumulated from regions and private persons with a higher level of quality of life. Thus, instead of adjusting the underlying indicators to stimulate investments of the prosumer in solar PV systems, the legislator should apply the result to allocate funds to regions and private households depicting a lower level of quality of life and, thus, help to increase the quality of life.

The implications from the second factor are quite important as urbanization world-wide is accelerating. As private households are mostly able to invest in solar PV systems if they own their own property, the high number of renters in cities poses a barrier for further investments. Therefore, the legislator must make sure that inhabitants that are renting a house or apartment can invest in the energy transition as well.

Until now, investments from prosumers in Germany mostly come from the south of Germany where the quality of life is high, and many inhabitants live in their own properties in rural areas. In contrast, inhabitants living in cities with a low level of freestanding houses and a lower level of property owners have been excluded from investments in solar PV systems. Thus, the increasing number of inhabitants accompanied by sharp increases in property prices states a huge barrier for investments of the prosumer. In order to overcome this barrier, the legislator should adjust the legislation regarding the opportunity of tenants to participate in the energy transition. Some German states like Hamburg announced making it mandatory for new buildings to integrate solar PV systems and to offer the tenants the possibility to participate.

The fourth factor could be named industrialized regions. In Germany, the most important industrialized regions are located in the south of Germany. Within this region, the combination of well-paid industrial jobs and rural areas with a high level of homeowners seem to be a driver for investments in solar PV systems from private households. Similar to factor one, the risk of a high level of cumulated investments in particular regions diminish the targets of the legislator to broaden the number of participants that take part in the energy transition. But, as the acceptance for this long-lasting project is quite important, the legislator should think about special programs for deindustrialized regions that have lost added value during previous years due to the phase-out of coal-fired power plants and the decrease in the coal, iron and steel industry.

Author Contributions: Conceptualization P.R. and M.S.; methodology P.R. and M.S.; software, M.S.; formal analysis, M.S.; investigation, P.R. and M.S.; resources, P.R. and M.S.; data curation, P.R. and M.S.; writing—original draft preparation, P.R. and M.S.; writing—review and editing, P.R. and M.S.; supervision, P.R. and M.S. All authors have read and agreed to the published version of the manuscript.

Funding: This research received no external funding.

Institutional Review Board Statement: Not applicable.

Informed Consent Statement: Not applicable.

Data Availability Statement: Multiple data sources have been used, specified in the Materials and Methods section.

Conflicts of Interest: The authors declare no conflict of interest.

References

1. Altmaier, P. Die Energiewende ist die größte umwelt- und wirtschaftspolitische Herausforderung zu Beginn des 21. Jahrhunderts. *Politische Bild.* **2013**, *46*, 7–24.
2. *Gesetz zur Reduzierung und zur Beendigung der Kohleverstromung und zur Änderung Weiterer Gesetze of 08.08.2020*; BGB I: Hongkong, China, 2020; p. 1818.
3. Frauenhofer ISE. Available online: https://www.ise.fraunhofer.de/content/dam/ise/en/documents/publications/studies/recent-facts-about-photovoltaics-in-germany.pdf (accessed on 28 September 2021).
4. *Gesetz für den Vorrang Erneuerbarer Energien (Erneuerbare-Energien-Gesetz-EEG) of 29.03.2000*; BGB I: Hongkong, China, 2000; p. 305.
5. Eising, M.; Hobbie, H.; Möst, D. Future wind and solar power market values in Germany—Evidence of spatial and technological dependencies? *Energy Econ.* **2020**, *86*, 104638. [CrossRef]
6. Statista GmbH. Available online: https://de.statista.com/statistik/daten/studie/29264/umfrage/neu-installierte-nennleistung-von-solarenergie-in-deutschland-seit-2004/ (accessed on 28 September 2021).
7. Bundesnetzagentur. Available online: https://www.bundesnetzagentur.de/DE/Sachgebiete/ElektrizitaetundGas/Unternehmen_Institutionen/ErneuerbareEnergien/ZahlenDatenInformationen/EEG_Registerdaten/EEG_Registerdaten_node.html;jsessionid=74CAD393E43422BAA66E0931D1F9DC05 (accessed on 28 September 2021).
8. Poier, S. Towards a psychology of solar energy: Analyzing the effects of the Big Five personality traits on household solar energy adoption in Germany. *Energy Res. Soc. Sci.* **2021**, *77*, 102087. [CrossRef]
9. Kempton, W.; Tomi´c, J. Vehicle-to-grid power implementation: From stabilizing the grid to supporting large-scale renewable energy. *J. Power Sour.* **2005**, *144*, 280–294. [CrossRef]
10. E. Moula, M.; Maula, J.; Hamdy, M.; Fang, T.; Jung, N.; Lahdelma, R. Researching social acceptability of renewable energy technologies in Finland. *Int. J. Sustain. Built Environ.* **2013**, *2*, 89–98. [CrossRef]
11. Wüstenhagen, R.; Bilharz, M. Green energy market development in Germany: Effective public policy and emerging customer demand. *Energy Policy* **2006**, *34*, 1681–1696. [CrossRef]
12. Bilgen, S.; Keleş, S.; Kaygusuz, A.; Sari, A.; Kaygusuz, K. Global warming and renewable energy sources for sustainable development: A case study in Turkey. *Renew. Sustain. Energy Rev.* **2008**, *12*, 372–396. [CrossRef]
13. Ince, D.; Vredenburg, H.; Liu, X. Drivers and inhibitors of renewable energy: A qualitative and quantitative study of the Caribbean. *Energy Policy* **2016**, *98*, 700–712. [CrossRef]
14. Dui´c, N.; da Graça Carvalho, M. Increasing renewable energy sources in island energy supply: Case study Porto Santo. *Renew. Sustain. Energy Rev.* **2004**, *8*, 383–399. [CrossRef]
15. Shahrestani, M.; Yao, R.; Luo, Z.; Turkbeyler, E.; Davies, H. A field study of urban microclimates in London. *Renew. Energy* **2015**, *73*, 3–9. [CrossRef]

16. Divic, V. Wind energy potential in the Adriatic coastal area, Croatia-field study. In *International Symposium on Computational Wind Engineering*; William and Ida Friday Center for Continuing Education: Chapel Hill, NC, USA, 2010.
17. Makki, A.A.; Mosly, I. Factors Affecting Public Willingness to Adopt Renewable Energy Technologies: An Exploratory Analysis. *Sustainability* **2020**, *12*, 845. [CrossRef]
18. He, Z.-X.; Xu, S.-C.; Li, Q.-B.; Zhao, B. Factors That Influence Renewable Energy Technological Innovation in China: A Dynamic Panel Approach. *Sustainability* **2018**, *10*, 124. [CrossRef]
19. Olanrewaju, B.T.; Olubusoye, O.E.; Adenikinju, A.; Akintande, O.J. A panel data analysis of renewable energy consumption in Africa. *Renew. Energy* **2019**, *140*, 668–679. [CrossRef]
20. Grzeszczyk, T.A.; Izdebski, W.; Izdebski, M.; Waściński, T. Socio-economic Factors Influencing the Development of Renewable Energy Production Sector in Poland. *E+M Èkon. Manag.* **2021**, *24*, 38–54. [CrossRef]
21. Fatima, N.; Li, Y.; Ahmad, M.; Jabeen, G.; Li, X. Factors influencing renewable energy generation development: A way to environmental sustainability. *Environ. Sci. Pollut. Res.* **2021**, *28*, 51714–51732. [CrossRef]
22. Williamson, O.E. Comparative Economic Organization: The Analysis of Discrete Structural Alternatives. *Adm. Sci. Q.* **1991**, *36*, 269. [CrossRef]
23. Rindfleisch, A.; Heide, J. Transaction Cost Analysis: Past, Present, and Future Applications. *J. Mark.* **1997**, *61*, 30–54. [CrossRef]
24. Bussolo, M.; Whalley, J. Exploring the links between transaction costs, income distribution and economic performance in a case study for Colombia. *Econ. Int.* **2003**, *94–95*, 235–260. [CrossRef]
25. Geetha, N.; Ramesh, M. A study on relevance of demographic factors in investment decisions. *Perspect. Innov. Econ. Bus.* **2012**, 14–27. [CrossRef]
26. Shanmugham, R.; Ramya, K. Impact of Social Factors on Individual Investors' Trading Behaviour. *Procedia Econ. Financ.* **2012**, *2*, 237–246. [CrossRef]
27. Ciobanu, R.; Bahna, M. The Social, Cultural and Political Factors that Influence the Level of Mergers and Acquisitions. *Int. J. Acad. Res. Account. Financ. Manag. Sci.* **2015**, *5*. [CrossRef]
28. Johnson, R.A.; Wichern, D.W. *Applied Multivariate Statistical Analysis*; Pearson: London, UK, 2014; Volume 6.
29. Yang, B. Factor analysis methods. In *Research in Organizations: Foundations and Methods of Inquiry*; Louisiana State University: Baton Rouge, LA, USA, 2005; pp. 181–199.
30. McNeish, D. Thanks coefficient alpha, we'll take it from here. *Psychol. Methods* **2018**, *23*, 412–433. [CrossRef] [PubMed]
31. Lorenzo-Seva, U. *How to Report the Percentage of Explained Common Variance in Exploratory Factor Analysis*; Department of Psychology: Tarragona, Italy, 2013.

Article

An Energy Consumption Approach to Estimate Air Emission Reductions in Container Shipping

Ernest Czermański [1,*], Giuseppe T. Cirella [2], Aneta Oniszczuk-Jastrząbek [1], Barbara Pawłowska [2] and Theo Notteboom [3,4,5,6]

1   Department of Maritime Transport and Seaborne Trade, Faculty of Economics, University of Gdansk, 81-824 Sopot, Poland; aneta.oniszczuk-jastrzabek@ug.edu.pl

2   Department of Transport Economics, Faculty of Economics, University of Gdansk, 81-824 Sopot, Poland; gt.cirella@ug.edu.pl (G.T.C.); barbara.pawlowska@ug.edu.pl (B.P.)

3   Center for Eurasian Maritime and Inland Logistics, China Institute of FTZ Supply Chain, Shanghai Maritime University, Shanghai 201306, China; theo.notteboom@ugent.be

4   Maritime Institute, Faculty of Law and Criminology, Gent University, B-9000 Gent, Belgium

5   Faculty of Business and Economics, University of Antwerp, 2000 Antwerp, Belgium

6   Antwerp Maritime Academy, 2000 Antwerp, Belgium

*   Correspondence: ernest.czermanski@ug.edu.pl; Tel.: +48-502-241-414

Abstract: Container shipping is the largest producer of emissions within the maritime shipping industry. Hence, measures have been designed and implemented to reduce ship emission levels. IMO's MARPOL Annex VI, with its future plan of applying Tier III requirements, the Energy Efficiency Design Index for new ships, and the Ship Energy Efficiency Management Plan for all ships. To assist policy formulation and follow-up, this study applies an energy consumption approach to estimate container ship emissions. The volumes of sulphur oxide (SO_x), nitrous oxide (NO_x), particulate matter (PM), and carbon dioxide (CO_2) emitted from container ships are estimated using 2018 datasets on container shipping and average vessel speed records generated via AIS. Furthermore, the estimated reductions in SO_x, NO_x, PM, and CO_2 are mapped for 2020. The empirical analysis demonstrates that the energy consumption approach is a valuable method to estimate ongoing emission reductions on a continuous basis and to fill data gaps where needed, as the latest worldwide container shipping emissions records date back to 2015. The presented analysis supports early-stage detection of environmental impacts in container shipping and helps to determine in which areas the greatest potential for emission reductions can be found.

Keywords: container shipping; emissions; maritime transport; sustainable shipping; green shipping; IMO

Citation: Czermański, E.; Cirella, G.T.; Oniszczuk-Jastrzębek, A.; Pawłowska, B.; Notteboom, T. An Energy Consumption Approach to Estimate Air Emission Reductions in Container Shipping. *Energies* **2021**, *14*, 278. https://doi.org/10.3390/en14020278

Received: 11 December 2020
Accepted: 4 January 2021
Published: 6 January 2021

Publisher's Note: MDPI stays neutral with regard to jurisdictional claims in published maps and institutional affiliations.

1. Introduction

The world's container shipping fleet (fully cellular) consisted of approximately 5600 vessels in 2018, representing only 8% of the total fleet tonnage in global shipping. However, in terms of sailing distance container ships are responsible for 17% of all maritime transport [1,2]. The container shipping fleet is diversely structured, both in terms of vessel size and cargo-carrying capacity, with eight principal sizes varying in numerical strength and capacity (Table 1). In December 2018, the container shipping fleet delivered a combined deadweight tonnage of 295,746,617 dwt. The average vessel was 224.2 m in length, 32.2 m in width, and had 11.2 m draught. The average technical speed for the entire fleet was 20.5 kn, while the average combined engine power was 28,089 kW, with the whole fleet amounting to 157,413,558 kW (see Supplementary Materials) [2]. The first four parameters impact a ship's energy requirement while the last two directly affect fuel consumption.

Table 1. Size and carrying capacity of the container shipping fleet by vessel type, 2018.

	Size		Carrying Capacity (TEU [1])	
Vessel Type	**Total Vessels**	**Fleet Percentage**	**Average**	**Standard Deviation**
Containership-Small Feeder	953	17.0	604.2	265.0
Containership-Regional Feeder	1393	24.9	1415.4	307.6
Containership-Feedermax	758	13.5	2530.7	241.9
Containership-Sub-Panamax	202	3.6	3368.3	246.8
Container-Baby post-Panamax	193	3.4	4362.8	526.7
Containership-Panamax	533	9.51	4508.3	335.0
Containership-Post-Panamax	925	16.5	7572.3	1382.4
Containership-ULCS [2]	647	11.6	14,543.0	3549.7
Total	5604	100.0	4426.3	4533.4

[1] twenty-foot equivalent unit. [2] ultra large container ship. Source: own compilation based on IHS Markit Portal [2].

According to data from 2015, container ships consumed 80 million metric tons of marine fuel, a figure amounting to 25% of total fuel consumption by all ships worldwide. To stress energy usage, container shipping represents 26% of the shipping industry's total energy intake [1]. As such, no other segment of maritime shipping can boast such extravagant figures [3]. Possible solutions to reduce fuel consumption include: (1) reducing ship speed, (2) installing auxiliary propulsion systems (e.g., kites), (3) streamlining the ship's hull (e.g., slippery layers on the submerged part of the hull) and (4) route optimization systems concerning navigation conditions, speed, heeling, and other voyage parameters to reduce fuel consumption throughout the voyage [4,5].

The International Maritime Organization (IMO) has estimated that the maritime shipping industry contributes 2.5 to 3.0% of annual human-produced carbon dioxide (CO_2) emissions in which the largest portion of that is derived from container shipping [1]. As such, container shipping is responsible for a significant part of the world's burning of fossil fuels and ocean pollution. Alongside CO_2, other greenhouse gas (GHG) emissions released by ships include SO_x, NO_x, and PM, which are highly toxic, create air pollution and cause acid rain (i.e., via irregular pH levels)—the health effects of pollutants on human health are often difficult to estimate as they depend on the substance as well as its concentration and exposure time. According to the World Health Organization (WHO) [6,7], approximately 1.3 million people die prematurely every year in cities as a result of urban air pollution. In Europe, according to the European Environmental Agency, there are 350,000 premature deaths due to over exposure to $PM_{2.5}$ and 20,000 premature deaths due to exposure to high O_3 concentrations [8,9]. The WHO has also shown that exposure to $PM_{2.5}$ accounts for 8% of lung cancer deaths, 5% of cardiovascular deaths, and 3% of respiratory infections worldwide [10]. It also indicates that an increased risk of morbidity and even mortality from respiratory diseases is associated with exposure to NO_2, also in concentrations below the limit values [11]. Another effect includes black carbon which reduces ice cover and overall absorption rates that create higher levels of heat due to positive radiative forces (e.g., climate change). The resulting short- and long-term effects on health are important reasons for limiting emission levels [12].

Due to the fact that container shipping is such a critical part of the global economy, ceasing container ship activity is not a feasible option [13]. Yet, it is essential that measures are implemented to first reduce and eventually halt emissions. To this end, some important steps have already been taken such as the judicial implementation of IMO's MARPOL Annex VI that binds limitations on the main air pollutants contained in ships' exhaust as well as further plans of implementing Tier III requirements), Energy Efficiency Design Index (EEDI) for new ships, and Ship Energy Efficiency Management Plan (SEEMP) for all ships [1,14].

To assist policy formulation and follow-up, this paper uses an energy consumption approach to determine air emissions generated by container shipping in 2018 and to simulate reductions in emissions for 2020. Using the baseline information on vessel size and carrying capacity of the container shipping fleet, we piece together the estimated volume of air emissions from 2016 to 2020 using cross-referenced [2] data. The calculations are supported by a mapping analysis which portrays a visual illustration of the estimated potential for reducing shipping emissions. In term of regulatory measures, this paper should be aligned with the IMO's MARPOL Annex VI for CO_2 [1], Regulation 14 for SO_x and PM [15] and Regulation 13 for NO_x [16]. To better understand the shipping container industry's connection with emission levels the state of the art of technologies, as well as fuel options and other alternatives, are examined. The calculations from this paper bare some significance as they give a sneak peek into future estimation of emission levels generated by specific shipping routes and shipping lines as well as determine the amount of emissions per unit of a container ship's transport work. This will, in turn, help to determine average emissions per container-mile (i.e., unit of transport work). Moreover, the presented analysis supports early-stage detection of environmental impacts in container shipping and helps to determine in which areas the greatest potential for emission reductions can be found.

The paper is structured as follows. After a brief discussion on possible methodological approaches for the measurement of ship emissions, we introduce the building blocks of the energy consumption approach and provide more details on the data collection and design aspects of the research and the mapping of results. Section 4 elaborates on the empirical results with a focus on the worldwide energy requirements for the container shipping fleet. These outcomes are used to estimate the potential for reducing SO_x, NO_x, PM, and CO_2 emissions by comparing the situation for 2018 with future estimates under global cap conditions.

2. Methodological Approaches to the Measurement of Ship Emissions

There is a vast literature on the measurement of ship emissions including scientific work and regulation-related documents for IMO and others. One of the first seminal studies among the related IMO documents was in 2000, in which an international consortium, led by Marintek, delivered the "Study of Greenhouse Gas Emissions from Ships" which included an estimation of ship emissions for 1996 and an examination of emission reduction possibilities. The two main methodological approaches can be adopted to measure ship emissions: a fuel-based approach (top-down) and an activity-based (bottom-up) approach [17]. The first approach relies on marine fuel consumption data and fuel-related emission factors [18] and is particularly useful in the case no detailed information is available on ship movements. The activity-based approach method requires very detailed datasets on the technical characteristics and operations of individual vessels. Based on these datasets, emissions of an individual ship can be calculated and aggregated to obtain fleet emission estimates [19]. Some researchers have followed hybrid approaches by combining the top-down and bottom-up approaches. While the bottom-up methodology in principle generates more fine-meshed results, it requires vast amounts of reliable vessel and traffic data [20]. Additionally, the activity-based approach often relies on the use of average input parameters like engine load factors, time spent in port, fuel consumption rate, and emission factors which greatly depend on the ship characteristics (e.g., age, size, fuel type) and the (changing) market conditions [21].

Corbett and Köhler [22], Endresen et al. [23], Eyring et al. [24] were among the early studies presenting detailed methodologies for constructing fuel-based inventories of ship emissions mainly based on engine power and vessel activity data. These methods to measuring ship emissions have been refined and adapted in later studies. Using fuel consumption as a main input, Psaraftis and Kontovas [25] presented an analysis on emissions of the world fleet for major ship types, i.e., bulk carriers, crude oil tankers, container vessels, product or chemical carriers, liquefied natural gas (LNG) carriers, liquefied petroleum gas carriers, reefer vessels, Ro-Ro vessels, and general cargo ships. The literature review study of Nunes et al. [17] analyzed 26 papers published between 2010 and 2016 using the activity-based methodology to estimate ship emissions. Most of the authors allocating emissions by ship type concluded that container ships were the main pollutant emitters. Cariou et al. [26] followed a mixed approach (bottom-up and top-down) by developing a model to estimate the total CO_2 emissions in container shipping per trade lane, using three building blocks: a trade-related port time module, a trade-related liner service module, and a vessel design speed fuel consumption module. The study reports an average CO_2 emission in 2016 of 58 g per TEU-km against 50 reported in a similar study by BSR Clean Cargo Working Group [27].

The fuel-based approach (top-down) and activity-based approach (bottom-up) have also been deployed to forecast ship emissions and associated environmental impacts. For example, Song and Shon [18] present a scenario-based analysis to predict ship emissions in the port of Busan for the medium and long-term (up to 2050), while Corbett et al. [28] produced forecasts for 2020, 2030, and 2050. Short-term vessel emission forecasts have been presented by Liu et al. [29] in the context of the Domestic Emission Control Areas in China. Liu and Duru [30] criticize the use of deterministic extrapolation and point estimates as a basis for emission forecasting and, therefore, propose a Bayesian algorithm approach which can generate a range of possible outcomes based on the probabilistic forecasting concept.

As mentioned earlier, this study applies an energy-consumption approach to determine air emissions generated by container shipping in 2018 and to simulate reductions in emissions for 2020. In view of determining the CO_2 emissions by global container shipping, this paper appends to the current worldwide CO_2 emission levels from shipping which dates back to 2007–2015 (i.e., Third IMO GHG Study from 2007 to 2012 and International Council on Clean Transportation's (ICCT) report from 2013 to 2015) (Table 2). The Fourth IMO GHG Study has been published in the last quarter of 2020 [31].

Table 2. Worldwide CO_2 emissions from international shipping as a total of global anthropogenic emissions, cargo carriage volume and transportation work performed, 2007–2015.

	Third IMO GHG Study						ICCT		
	2007	2008	2009	2010	2011	2012	2013	2014	2015
Global CO_2 emissions [1]	31,959	32,133	31,822	33,661	34,726	34,968	35,672	36,084	36,062
Total shipping [1]	1100	1135	977	914	1021	942	910	930	932
International shipping [1]	881	916	858	773	853	805	801	813	812
Cabotage [1]	133	139	75	83	110	87	73	78	78
Fisheries [1]	86	80	44	58	58	51	36	39	42
Share of shipping in global CO_2 emissions	3.4%	3.5%	3.1%	2.7%	2.9%	2.6%	2.5%	2.6%	2.6%
Cargo carriage [2]	8034	8229	7858	8409	8785	9197	9514	9843	10,024
Transportation work [3]	41,177	42,167	40,308	44,603	46,834	48,937	50,490	52,627	53,476
Correlation factor (E^T)	0.277	0.269	0.242	0.205	0.218	0.193	0.180	0.177	0.174

[1] million metric tons/year; [2] volume calculation is done using the STEAM3 model sourced from [3]; [3] million metric tons/Nm; correlation coefficient, i.e., E^T, represents the ratio between the amount of CO_2 emissions from shipping in tons of CO_2 and amount of annual transportation work carried out by the world fleet in tons of CO_2. Source: authors own elaboration based on Comer et al. [32] and statistical data from UNCTAD/RMT/2018 [33].

3. Methodology

3.1. Methodological Scope of the Energy Consumption Approach

In this paper, we present an energy consumption approach to the measurement of ship emissions of the container shipping fleet, i.e., a conceptual alternative energy approach, to measure worldwide energy requirements for container shipping. This approach considers the size of the fleet and parameters related to the required energy to operate this fleet. As indicated earlier, the application of the energy consumption approach can be utilized as a tool-based method when limited or no data records on detailed container ship movements are made available. It allows for the estimation of ship fuel consumption and emitted air emissions in the course of ship operations. A compilation of relevant air emissions data has been developed based on the various stages during the round voyage of a ship, i.e., total berth time, maneuvering within the port, entry into port, anchoring and—most importantly—the actual voyage at sea. This approach enables for a more thorough analysis of the environmental impact of shipping and helps to determine in which areas action can be taken to reduce air emissions and where the greatest potential for such reductions lie.

3.2. Data Collection, Design, and Variables

A database on container ships was created specifically for this research, with data for December 2018. The base dataset was purchased from the IHS [2] Maritime Portal. The data focuses on parameters crucial for estimating emissions and correlating the amount of fuel consumed by ships. Fuel consumption has been determined based on an assigned value to each ship in relation to its cargo-carrying capacity. Using this baseline method, a container ship with a carrying capacity of 1000 TEU, tested within the speed range of 10–17 kn, corresponds to Froude numbers ranging between 0.16 to 0.28 (Figure 1) [34–36]. It should be emphasized that the shape of the identified graph is cross-compatible with larger vessels without incurring a risk of methodological error. The calculation shows a steep increase in fuel consumption as speed increases. Systematically, at 10 kn the daily fuel consumption reaches 4.8 metric tons, at 15 kn (i.e., a 50% increase) it is 14.4 metric tons (i.e., 300% more), while at 17 kn (i.e., 13% up from 15 kn) the daily fuel consumption reaches 33.6 metric tons (i.e., 233% on top of the preceding amount). These calculations indirectly point to a strong increase in the levels of pollutants into the air as the ship speed increases.

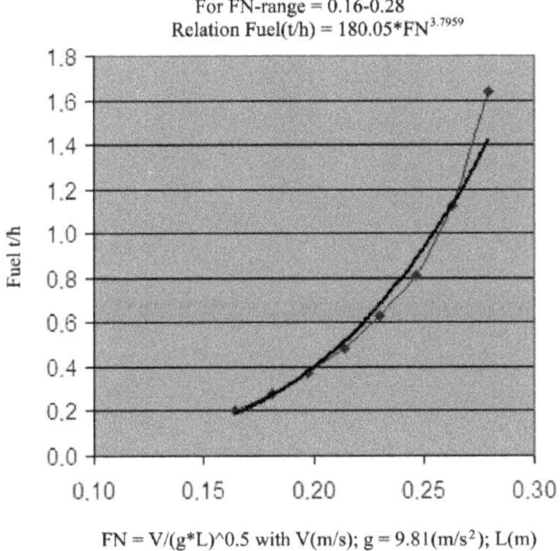

Figure 1. An example function illustrating fuel consumption relative to the ship's speed expressed as a Froude number.

To translate the results from energy use to emissions, the calculated fuel consumption figures have been correlated with emission level indices for each of the specific fuel types. The most widespread indicators link air emissions to energy density by using the following fuel types: heavy fuel oil (HFO) (i.e., 40.0 MJ/kg of fuel), marine diesel oil (MDO) (i.e., 42.7 MJ/kg) and LNG (i.e., 50.0 MJ/kg). The indices were converted in such a manner as to translate consumption of a given fuel (i.e., in metric tons) into emissions produced in the process of consumption (i.e., in metric tons of CO_2 and kilograms for other pollutants).

The results are presented in Table 3. The calculations identify the total air emissions for basic pollutant compounds derived from container shipping.

Table 3. Emissivity indices for selected marine fuels.

		[kg/t of Fuel]				Energy Density
Item	Fuel	CO_2	SO_X	NO_X	$PM_{2.5}$	Relative to MDO (%)
1	MDO 0.5%	3206.00	10.50	50.50	2.30	100
2	HFO 1.5%	3114.00	31.50	51.00	3.40	94
3	HFO 2%	3114.00	42.00	51.00	3.40	94
4	HFO 3.5%	3114.00	71.50	51.00	3.40	94
5	LSHFO 0.5%	3151.00	10.50	51.00	2.30	94
6	LSMGO 0.1%	3151.00	2.10	50.50	2.30	100
7	LNG	2750.00	<0.02	8.40	0.02	1170
8	Methanol	1375.00	0.00	26.10	0.02	46.8
9	HFO + SCRUBBER + SCR [1]	3176.28	0.84	7.65	0.51	94

[1] selective catalytic reduction. Source: authors own elaboration based on the assumptions of the Med Atlantic Ecobonus (MAE) Project, MAE External Cost Calculator Tool [37].

ICCT was consulted to determine the structure of marine fuels used in container shipping [32]. The predominant fuel used for the dataset is HFO whose percentage amounts to 93%, with the remaining 7% referring to distillates (Figure 2).

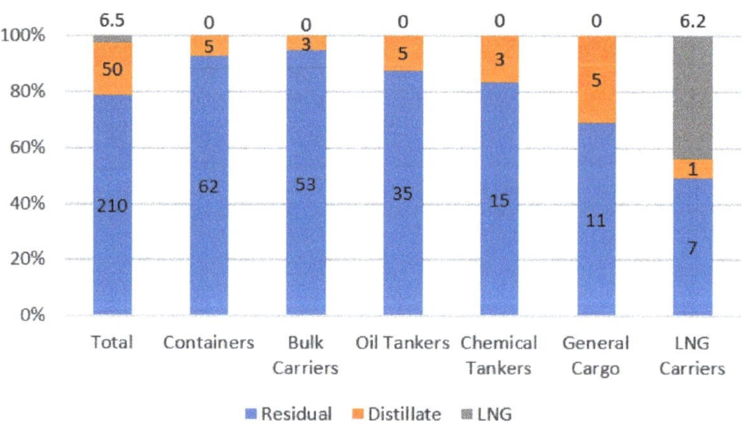

Figure 2. Classification by type of marine fuel according to the shipping sector, adapted from Comer et al. [32].

LNG accounts for a marginal share in the analyzed dataset, considering that the entire fleet had only four vessels powered by this type of fuel. Such a low share of LNG, within the overall fuel consumption figure, calls for a thorough analysis in order to explore the possibility of modernizing the fleet with a view to reducing overall emission levels.

3.3. Mapping

The mapping analysis portrays a visual illustration of the estimated potential for reducing shipping emissions. Then, two software packages were used for Geographic Information Systems (GIS) imagery and overlay supply, i.e., software Esri ArcGIS (version 10.7) and software Paint.Net (version 4.1.6). The GIS mapping calculations used the mean total for emission levels (i.e., SO_X, NO_X, PM, and CO_2) to deduct results from Equation (1).

$$R_r = delta(X_v) / X(V_{n0}) \tag{1}$$

where: R_r = reduction rate (mean), $delta(X_v)$ = difference in existing emission level (X_0) versus predicted emission level (X_1) by specific speed "V", "n" = speed between 14 and 18 kn, $X(V_{n0})$ = existing emission volume for specific "n" speed "V".

We apply the energy consumption approach to calculate the consumed energy (i.e., fuel) from container ships on a global scale. Due to a lack of records, automatic identification system (AIS) datasets (i.e., for all container ships worldwide) were single-handedly calculated to obtain real fuel consumption levels [2]. This method has not been used for the whole of the container shipping sector and provides first account and valuable intermediate results before the Fourth IMO GHG Study is published in the last quarter of 2020 [31].

3.4. Empirical Results

In a first step, the results focus on the worldwide energy requirements for the container shipping fleet. The fuel consumption type is determined via fuel use patterns and maximum continuous rating (MCR). Then, we identify the estimated potential for reducing SO_x, NO_x, PM, and CO_2 in four subsequent subsections. For each of these subsections emissivity is generated in 2018, future estimates are presented (i.e., under global capped conditions) and mapped results for 2020 are presented using the energy consumption approach.

3.4.1. Energy Requirements in Container Shipping

The world's container shipping fleet, which included some 5600 vessels in 2018, represents 149,483 MW of power available from auxiliary engines and power generators, translating into an energy requirement of 179 GW. When assuming operations at maximum performance (i.e., MCR = 1.0) annually (i.e., per 365 days) and an average requirement for HFO of 180 g/1kWh, the maximum fuel requirement is estimated to be 282,000,000 metric tons [1]. If we consider the average sailing time only (i.e., circa 250 days), the figure drops to around 193,000,000 metric tons. This is the actual fuel consumption figure for the entire global container shipping fleet. According to data from 2012, fuel consumption stood at 276,000,000 metric tons, of which 195,000,000 represented HFO and 81,000,000 MGO, respectively [1–3,19,38].

In order to calculate the annual emissions produced by container shipping, the average power output use is reduced to MCR = 0.85. This is, in fact, the power output level adopted for this research in which periodic change could occur depending on a ship owner's strategy and current transport needs. Generally, however, this power output factor is valid, as vessels do not operate at maximum power due to the excessive fuel consumption and elevated risk of emergency this would bring. According to Comer et al. [32], the volume of fuel consumed in 2015 in the container shipping industry amounted to 66,860,000 metric tons, representing 25% of the combined use of all types of fuel in the world [39,40]. As noted, 93% of that figure accounts for HFO residual fuel (i.e., 30% of global use) and 7% distillates (i.e., 10%). In the case of HFO fuel usage, container shipping is the largest consumer in the entire shipping industry, while distillates are predominately used in ferry and ro-ro shipping. LNG consumption was measured at around 3000 metric tons which is a negligible value compared to the overall 6,200,000 metric tons consumed by gas carriers.

Fuel consumption was analyzed by type of shipping to determine fuel consumption patterns at the various stages of a voyage. A voyage consists of four stages: (1) moored at berth for cargo handling operations (including idle time), (2) maneuvering (i.e., at the port), (3) anchoring (i.e., before entering the port and while awaiting further instructions) and (4) time at sea (i.e., sailing time) [14,32]. Container ships use a mere 8% of fuel outside of the time spent at sea, as shown in Figure 3. This means the primary focus area for measures to reduce air emission levels should be when ships are sailing at sea. The ship owner usually has only partial control over the length of the anchoring stage, as it depends on congestion within the port and access-preventing weather conditions. Coordination between route optimization systems used by the ship operator and vessel traffic management systems (VTMS) used to manage port access routes can help to avoid any waiting time".

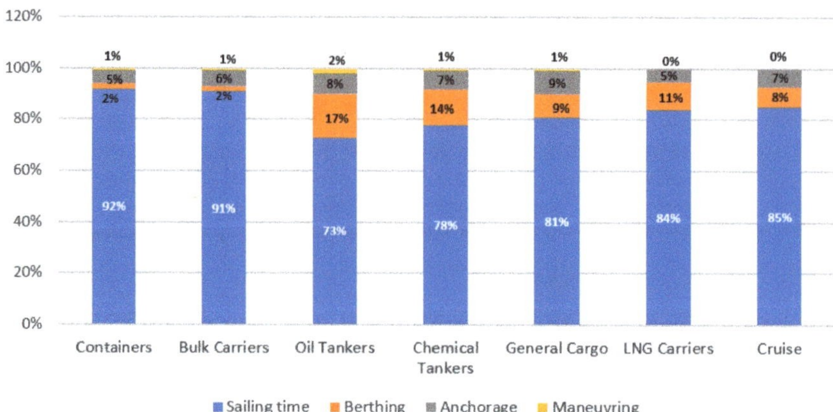

Figure 3. Energy consumed by ships according to the stage of voyage and type of fleet, adapted from Comer et al. [32].

Remarkably, container shipping is characterized by fuel use disproportions across the various stages of the sea voyage. According to Comer et al.'s [32] survey, as shown in Figure 3, cargo handling time in port is kept to a minimum as a result of the pressure exerted by shipping lines on terminal operators to reach a high terminal productivity in terms of the amount of cargo loaded and discharged per time unit.

To determine the fuel volume consumed by container ships in 2018, the technical parameters of ships, relating to engine power, cargo-carrying capacity in TEU and length were examined. Table 4 presents the calculated totals. The volume of fuel consumed in the container shipping industry was estimated at 117,800,000 metric tons in 2018. It was assumed for the purpose of the calculations that this figure is a theoretical value corresponding to 17 kn, i.e., a speed level approximating the average commercial speed reported by Marine Traffic [41] and confirmed by the authors in trial measurements executed in 2018.

Table 4. Calculated fuel consumption figures for container ships in 2018.

V [kn]	C_{MGO} [t]	C_{HFO} [t]	$\Sigma C_{(MGO+HFO)}$ [t]
14	6,821,449	90,627,824	97,449,273
15	7,333,524	97,032,536	104,366,060
16	7,784,678	103,425,006	111,209,684
17	8,247,186	109,569,756	117,816,942
18	9,852,961	130,903,629	140,756,590

Source: own calculations based on data from IHS Markit Portal [2].

To fully illustrate the relationship between fuel consumption and speed, Table 4 sets out theoretical values for other speeds within the range of 14–18 kn. The obtained results will be used to calculate emissivity to air for the most important compounds such as SO_x, NO_x, PM, and CO_2.

3.4.2. Estimated Potential for Reducing SO_x Emissions

Container shipping stands out as using a very large share of HFO-type residual fuel with various levels of leftover sulphur content. After establishing Sulphur Emission Control Area (SECA) zones, the fuel used in ship propulsion systems was 3.5%. Prior to 1 January 2015, low-sulphur 1.5% HFO fuel was used. Afterwards, ship owners switched to MGO compliant with the 0.1% requirement in which the use of scrubbers is minimal and no conversions to LNG were noticed.

As a result, it would be correct to say that the previously large demand for 1.5% HFO fuel has disappeared or declined considerably. A direct consequence of this is the reduced emissivity of sulphur oxides. According to calculations made from our approach,

7,850,000 metric tons of sulphur compounds were emitted in 2018 in container shipping alone. This is an approximate number that can be verified by tracking ships in motion at sea with the AIS and locating their position relative to the SECA zone, which can be challenging for future research projects [20,42]. To illustrate the changes that can occur as a result of variations in ship speed, Table 5 shows the calculated SO_x emissions for the speed range of 14 to 18 kn. It is evident a clear positive, though slightly regressive, correlation depicts SO_x emissivity rises, although at decreasing rates, in step with increasing speed.

Table 5. Estimated figures for SO_x emissivity generated by container shipping in 2018.

V [kn]	SO_x MGO [t]	SO_x HFO [t]	$\sum SO_x$ [t]
14	14,325	6,479,889	6,494,214
15	15,400	6,937,826	6,953,227
16	16,348	7,394,888	7,411,236
17	17,319	7,834,238	7,851,557
18	20,691	9,359,609	9,380,301

Source: own calculations based on data from IHS Markit Portal [2].

These calculations can be supplemented with an attempt to determine SO_x emissions following the entry into force of the so-called global cap in 2020. The calculations used emissivity coefficients for 0.5% fuel which has become the most widespread standard in the face of the existing cap. It was concluded based on an analysis of how many ships on order, whose delivery date falls after 2019, were equipped with scrubbers. The total number of these ships will reach 190 after 2020. Table 6 illustrates their percentage within the entire existing fleet and the number of ordered ships in total.

Table 6. Ordered container ships to be delivered in 2019–2022.

Year	Number of Scrubber-Equipped Ships	Total	Scrubber-Equipped Ships (%)	Total Fleet (%)
2019	26	160	16.25	2.24
2020	53	155	34.20	3.13
2021	16	39	41.00	3.40
2022	0	3	0.00	3.39
Total	95	403	23.60	–
Container ship fleet in operations in 2018				
Until 2018	95	5238	0.17	1.71

Source: authors own elaboration based on IHS Markit Portal [2].

There were 24 LNG-powered ships on order in late 2018 (i.e., this total does not include 2 ships whose delivery, scheduled for 2018, could not be confirmed). The data indicates LNG is not an alternative for 0.5% HFO since emissions were still found to exist at significant levels; however, other assumptions remained as they were. Unsurprisingly, SO_x emissions will face a drastic reduction by more than 85%, translating quantitatively to 6,600,000 metric tons for HFO alone (Table 7). Using GIS, annual SO_x emissions for 2020 visually illustrate a change from high to low levels with the introduction of the energy consumption approach (Figure 4).

A world-scale limit of sulphur content set at 0.1% requires either a full transition to 0.1% MGO fuel, installation of scrubbers, or using another solution (i.e., the latter is not included in the calculations). Such a development allows for an even greater reduction of SO_x which has been linked as the precursor to acid rain and atmospheric particulates [43]. In other words, a global limit of 0.1% (i.e., implemented via worldwide SECA regulation) minimizes sulphur emissions which amount to 230,000 metric tons annually. This signals a significant step in maneuvering toward zero-emissivity of SO_x in container shipping.

Table 7. Estimated amount of SO$_x$ emissivity in container shipping under global cap conditions (i.e., 0.5% HFO fuel) for 2020.

V [kn]	SO$_x$ MGO [t]	SO$_x$ HFO [t]	$\sum$ SO$_x$ [t]
14	14,325	951,592	965,917
15	15,400	1,018,842	1,034,2420
16	16,348	1,085,963	1,102,310
17	17,319	1,150,482	1,167,801
18	20,691	1,374,488	1,395,179

Source: based on extrapolated data from IHS Markit Portal [2].

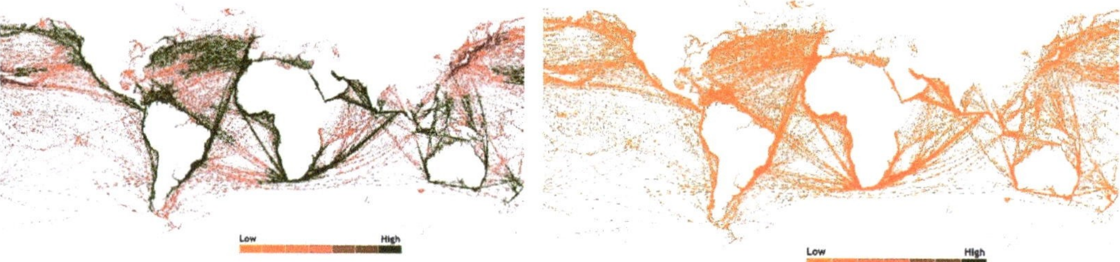

Figure 4. Annual global container shipping SO$_x$ emission levels for 2020, (left) base situation for 2018 and (right) estimates using the energy consumption approach, based on extrapolated data from IHS Markit Portal [2].

3.4.3. Estimated Potential for Reducing NO$_x$ Emissions

Currently, the use of NO$_x$ emissions, derived from fuels used in container shipping, does not have any major reduction plans with exception of the implementation of the Tier III limits. According to results obtained for our approach, NO$_x$ emissions for 2018 amounted to 6,000,000 metric tons (Table 8). The results indicate the determined emissions would grow in step with increasing speed, however, at a progressively slower rate.

Table 8. Estimated amount of NO$_x$ emissions generated by container shipping in 2018.

V [kn]	NO$_x$ MGO [t]	NO$_x$ HFO [t]	$\sum$ NO$_x$ [t]
14	344,483	4,622,019	4,966,502
15	370,343	4,948,659	5,319,002
16	393,126	5,274,675	5,667,801
17	416,483	5,588,058	6,004,540
18	497,574	6,676,085	7,173,660

Source: based on extrapolated data from IHS Markit Portal [2].

The limit for sulphur content in marine fuel at 0.5% since 2020 does not affect the volume of NO$_x$ emissions. Our approach has shown that substituting 0.5% HFO fuel for 3.5% HFO has no effect on NO$_x$ emissivity. However, crucial changes will come in the wake of the full implementation of the Tier III conditions following 2021. This approach uses NO$_x$ emissions reduction parameters set for SCR installations at 7.65 g per metric ton of fuel, irrespective of the fuel type. SCR purifies ship engine exhaust gases no matter where it is installed. Assuming that the Tier III would extend to all sea areas, the resulting NO$_x$ emissions reduction could reach 79.1%, translating to a reduction of 4,749,849 metric tons annually compared with the existing emissions (Table 9). For 2020, the GIS analysis for NO$_x$ emissions show a change from high to low levels with the introduction of the energy consumption approach (Figure 5).

Table 9. Estimated amount of NO_x emissions in container shipping under Tier III conditions (i.e., with SCR used) for 2021.

V [kn]	NO_x MGO [t]	NO_x HFO [t]	$\sum NO_x$ [t]
14	52,184	693,303	1,037,786
15	56,101	742,299	1,112,642
16	59,553	791,201	1,184,327
17	63,091	838,209	1,254,691
18	75,375	1,001,413	1,498,987

Source: own calculation based on data from IHS Markit Portal [2].

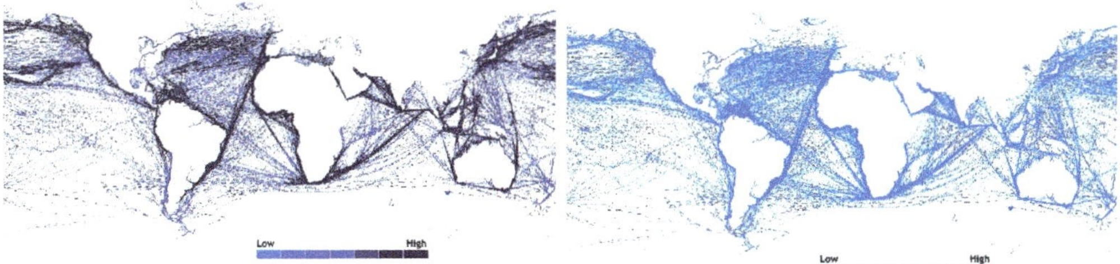

Figure 5. Annual global container shipping NO_x emission levels for 2020, (left) base situation for 2018 and (right) estimates using the energy consumption approach, based on extrapolated data from IHS Markit Portal [2].

As such, it is worth noting that possible efforts to reduce ship-derived NO_x emissions end there, as no other traditional fuel generates less emissions than this compound. Only possible alternatives beyond nitrogen oxide fuel would be the introduction of renewable sources such as hydrogen or electrically powered propulsion systems in which the sourcing process of these fuels (i.e., including electricity generated batteries or liquid hydrogen production) would equate to zero-emissivity [44].

3.4.4. Estimated Potential for Reducing PM Emissions

In quantitative terms, ship-derived particulate emissions seem to put a smaller strain on the environment. As shown in Table 10, these emissions in container shipping could amount to around 391.5 metric tons in 2018. However, we should keep in mind the strongly negative effect of PM on human health and its contribution to icecap melting around the world. Just as with other compounds, the size of PM emissions is growing in step with the increase in ship speed and fuel consumption. Additionally, just as in the previous cases, the increases are degressive.

Table 10. Estimated amount of PM emissions generated by container shipping in 2018.

V [kn]	PM MGO [t]	PM HFO [t]	$\sum$ PM [t]
14	15,689	308,135	323,824
15	16,867	329,911	346,778
16	17,905	351,645	369,550
17	18,968	372,537	391,506
18	22,662	445,072	467,734

Source: own elaboration based on data from IHS Markit Portal [2].

As the kind of fuel used by ships has a considerable effect on the volume of PM emissions, the calculation adopted a set global limit of 0.5% for 2020. The results of the calculations are presented in Table 11, suggesting that the transition to 0.5% HFO will enhance the reduction in PM emissions by 30.8%. In quantitative terms, under reference conditions, this would result in a reduction of more than 120,000 metric tons. The GIS analysis for PM emission levels in 2020 illustrate the change from high to medium with the introduction of the energy consumption approach (Figure 6).

Table 11. Estimated amount of PM emissions in container shipping under global cap conditions (i.e., 0.5% HFO fuel) for 2020.

V [kn]	PM MGO [t]	PM HFO [t]	$\sum$ PM [t]
14	15,689	208,444	224,133
15	16,867	223,175	240,042
16	17,905	237,877	255,782
17	18,968	252,010	270,979
18	22,662	301,078	323,740

Source: own calculations based on data from IHS Markit Portal [2].

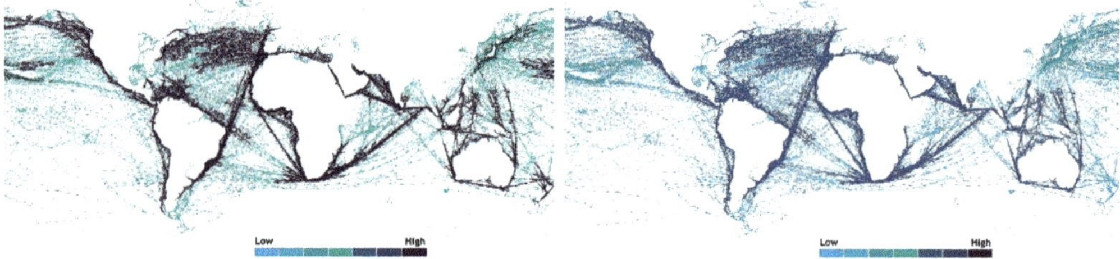

Figure 6. Annual global container shipping PM emission levels for 2020, (left) base situation for 2018 and (right) estimates using the energy consumption approach, based on extrapolated data from IHS Markit Portal [2].

It should be noted that further reductions in PM emissions are possible. Firstly, a shift towards LNG- or methanol-powered engines will bring about a radical decrease in these emissions to 2000 metric tons annually, a desirable step towards zero-emissivity, as well as, as previously discussed, the switch and implementation of hydrogen fuel or electrical propulsion systems.

3.4.5. Estimated Potential for Reducing CO_2 Emissions

The most thorough survey of CO_2 emissions is the 2017 ICCT report based on 2015 data. Besides SO_x, NO_x, PM and CO_2, the analysis also looked at CO, BC, N_2O and CH_4 emissions. The analysis of CO_2 emissions in maritime shipping shows that its volume in 2015 came close to 1,000,000,000 metric tons. The container shipping industry, which operates a relatively small number of ships, was responsible for the largest portion of that figure amounting to 23%, or a 4-point percent more than the bulk carrier fleet which is twice as large in number [32]. Representing the most important area of ship-derived emissions, emitted CO_2 was calculated using our approach in Table 12. The relational importance of CO_2 emissions is due to it being the largest in volume, amounting to 367,200,000 metric tons annually, and primarily centered on formalizing emissions regulations over the next three decades. The main factor affecting the volume of CO_2 emissions is the emissivity coefficient defining a ratio of 3.1 metric tons of CO_2 emissions per 1 metric ton of fuel consumed. Another concern is the fact that MGO, though more environmentally friendly in terms of sulphur and PM emissions, generates more CO_2 emissions than poor-quality residual fuel (i.e., these increases are shown in quantitative terms in Table 13). Incrementally, the data indicates this type of emission may reach 371,200,000 metric tons annually, exceeding the 2010 limit by more than 4,000,000 metric tons (Table 14), amounting to 1.1% of total CO_2 emissions. Our research indicates that setting limits to sulphur oxide emissions is counterproductive and leads to a corresponding increase in CO_2 emissions. In this context it is worth to recommend paying special attention always when new regulations and restrictions will be issued. The analyzed case shows that limiting one aspect can have a negative impact in another. A holistic approach is therefore, at most, essential.

Table 12. Estimated amount of CO_2 emissions generated by container shipping in 2018.

V [kn]	CO_2 MGO [t]	CO_2 HFO [t]	$\sum CO_2$ [t]
14	21,494,386	282,215,044	303,709,430
15	23,107,934	302,159,317	325,267,251
16	24,529,520	322,065,469	346,594,989
17	25,986,883	341,200,220	367,187,103
18	31,046,680	407,633,901	438,680,581

Source: own calculations based on data from IHS Markit Portal [2].

Table 13. Estimated amount of CO_2 emissions in container shipping under global cap conditions (i.e., 0.5% HFO fuel) for 2020.

V [kn]	CO_2 MGO [t]	CO_2 HFO [t]	$\sum CO_2$ [t]
14	21,494,386	285,568,273	307,062,659
15	23,107,934	305,749,521	328,857,455
16	24,529,520	325,892,194	350,421,714
17	25,986,883	345,254,301	371,241,184
18	31,046,680	412,477,335	443,524,015

Source: own calculations based on data from IHS Markit Portal [2].

Table 14. Changes in estimated amounts of CO_2 emissions generated by container shipping under global cap conditions as compared to 2018.

V [kn]	CO_2 MGO [t]	CO_2 HFO [t]	$\sum CO_2$ [t]
14	0.0	−3,353,229	−3,353,229
15	0.0	−3,590,204	−3,590,204
16	0.0	−3,826,725	−3,826,725
17	0.0	−4,054,081	−4,054,081
18	0.0	−4,843,434	−4,843,434

Source: own calculations based on data from IHS Markit Portal [2].

4. Discussion

We have examined the 2018 global container shipping fleet using the energy consumption approach with predictions for 2020. The findings are useful especially during interim breaks when the IMO or other international bodies have not released up to date and accurate information on emission levels in container shipping. A number of energy-related solutions to energy efficiency are worth noting. First, possible alternative propulsion via liquefied hydrogen fuels or electricity can prove to be very efficient [45–50]. A middle-of-the-road solution may also include switching to methanol. However, this is not practical due to limited output of this fuel globally as well as considerable costs related in converting propulsion systems. Second, viability may be supplied by way of new ship propulsion technology that will be introduced in response to the EEDI increasing energy efficiency standards. This can be seen clearly in the analysis of CO_2 emitted by container ships expressed in grams per ton-mile and according to vessel age, a correlation illustrated by the dispersed method in Figure 7. It illustrates dominant shifts towards larger TEU capacity and analyses concerned mainly with new ships (i.e., up to five years old). It is in this particular group of ships that the lowest rate of CO_2 emissions per unit of transport work exists. The future for air emissions will directly correspond with the EEDI standards which will operate in five-year cycles to reduce CO_2 emissions on new ships only, as well as the SEEMP for all ships. The EEDI is likely to become the most reliable driving factor behind shipowners' decisions, especially those operating container ships, to reduce emissions. Following 2025, these shipowners may be subject to pressure from technological restrictions. This follows the fact that traditional fuels and propulsion systems, including LNG, equipped with additional installations will no longer be enough to comply with the currently applicable limit on CO_2 emissions. The propulsion systems acceptable will be the ones powered by hydrogen, electricity, and methanol obtained from renewable sources.

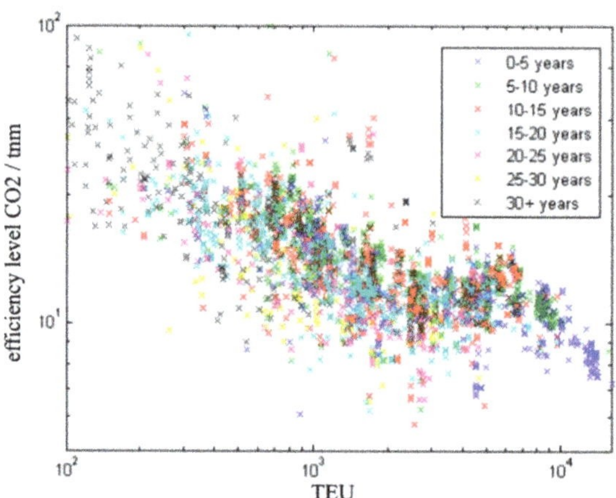

Figure 7. Container ship size correlated with emitted CO_2 (i.e., efficiency level gCO_2/tnm) per unit of transport work against TEU capacity, according to vessel age, adapted from Smith et al. [51].

In terms of the calculations relating to ship speeds, the used speed range covers 14 to 18 kn. If the speed of 17 kn is adopted as a reference point, we find the total annual fuel consumption in 2018 at 117,800,000 metric tons. The emission levels for this equates to over 7,850,000 metric tons of SO_x, 6,000,000 metric tons of NO_x, 391,000 metric tons of PM and 367,200,000 metric tons of CO_2. These figures illustrate that air emissions generated by global container shipping are significant. The latest data utilized from 2015 on CO_2 emissions amounts to just below 1,000,000,000 metric tons. The authors have determined the potential extent of reducing principal pollutants and air emissions after 2020 (i.e., in conjunction with global cap requisites) as a standard in upholding the United Nations Sustainable Development Goal 12 that ensures sustainable, responsible consumption and production [52]. Notably, the total emissions place container shipping as one of the top ranked emission activities (i.e., after electricity generation and industrial production). A further analysis of the data shows that the expected reduction in SO_x may reach an elevated 85%, while for PM a 33% reduction is estimated. Following 2021, the expected reduction in NO_x (i.e., when the Tier III limit is fully implemented) has been estimated at around 80%. As for CO_2 emissions, both the discussed limits will paradoxically swell emissions by around 4,000,000 metric tons annually as a result of transition to LSMGO fuel and increased demand for energy from scrubbers and SCR.

We should also consider a scenario of amending IMO regulations relating to EEDI by considering any innovative propulsion systems that may be developed, which will not only comply with the latest emissions limits but also—and most importantly—be affordable for ship owners. To best piece together best practices, the IMO's SEEMP measures and controls harmful air emissions from ships. Its implementation in coordination with EEDI tools can properly check ship pollution and act as an international baseline for managing ship energy efficiency, as well as the habits and know-how of the crew. Advanced features for implementation include new ship design and modification, as well as existing and older ship upgrading. Example amendments consist of innovation in hull shape, improved paints and antifouling, and various other techniques for wake equalization and flow separation. In terms of existing ships, to ensure energy efficiency is maximized they should maintain: (1) generators within an energy power management system in which they are run on as high load as possible when burning heavy fuel; (2) the stoichiometric ratio (i.e., air-fuel mixture); (3) high priority maintenance of fuel system to safeguard correct fuel viscosity before injectors, scavenge pressure control, boiler and economizer and viscotherm; and (4) performance of the main engine. As such, the combination of planning, implementation, monitoring, self-evaluation and improvement, speed optimization, weather routing, hull monitoring and maintenance, efficient cargo operation, and electric power management all contribute to optimizing best practices and overall ship efficiency. Ship efficiency enacts an important part of sustainable thinking in combination with fuel air emission data and research.

5. Conclusions

This study presented an energy consumption approach to determine the annual rate of air emissions generated by container shipping. Data energy projections have been updated for 2020 taking into account changes in energy choices in the wake of the implementation of the global cap in early 2020. The estimation of the fuel consumption level depends directly on ship speed which varies considerably throughout a ship's voyage. This issue is paramount in the reliability of the presented approach since emission level measures are at its highest at sea (i.e., not waiting at the quay, maneuvering within port, entering into port or anchoring). As such, it is necessary to adopt average values for the various fuels (i.e., HFO, MGO, and—marginally—LNG). This approach adds to the knowledge base for understanding on how worldwide container shipping can be optimized for such reductions. In line with IMO's initiatives in implementing MARPOL Annex VI (i.e., the limitation on air pollutants from ship exhaust) as well as impending Tier III conditions and the EEDI requirements for new ships. It is paramount to develop sustainable, reliable and state-of-the-art transport systems based on quality and resilient design. Further research would be beneficial if the IMO or any other international organization fails to update the latest worldwide container shipping records dating back to 2015.

The demand for container shipping is likely to increase in the long term, despite the negative effect on demand of major shocks such as the financial-economic crisis of 2009 and the (ongoing) COVID-19 crisis. The container box has become an essential part of the world economy and associated global supply chains [53]. To secure its future 'license to operate', the container shipping industry is challenged to measure emission levels and plan for their reduction to aid in decreasing air pollution and related side effects. The energy consumption approach applied to container shipping provides estimation-based figures for a sound understanding of current emission levels. To further develop an exact real-time record, sensor-based devices can be implemented and cross-integrated into the AIS. This practice would add an additional level of precision which could be optimized from a centralized hub or online source. From an analytical viewpoint, early-stage detection and continuous monitoring can help to better understand environmental impacts and to determine what areas offer the greatest potential for emission level reductions. The reduction in the ecological footprint of shipping is urgently needed in order to contribute to existing sustainable development goals. Within the domain of further reducing air emission levels in container shipping, voluntary declaration and the EEDI are valuable approaches that entail additional measures. The energy consumption approach is valuable to estimate ongoing emission reductions on a continuous basis when no other updated figures are available. The followed approach fills in the current data gap, as the latest worldwide container shipping emissions records date back to 2015.

Supplementary Materials: All the datasets originate from the IHS Markit Portal [2] website. The datasets generated and analyzed during the current study are available for purchase from their website at https://maritime.ihs.com/EntitlementPortal/Home/Index.

Author Contributions: Conceptualization, E.C.; methodology, E.C. and B.P.; formal analysis, E.C.; investigation, E.C.; resources, E.C., G.T.C., B.P., A.O.-J., T.N.; data curation, E.C., G.T.C.; writing—original draft preparation, E.C., A.O.-J., B.P., G.T.C., T.N.; writing—review and editing, B.P., G.T.C.,

T.N.; visualization and mapping, G.T.C.; supervision, T.N.; funding acquisition, E.C. All authors have read and agreed to the published version of the manuscript.

Funding: This research received no external funding.

Institutional Review Board Statement: Not applicable.

Informed Consent Statement: Not applicable.

Data Availability Statement: The datasets generated and analyzed during the study are not publicly available due to IHS Markit Portal [2] copyright but are available from the corresponding author on reasonable request.

Acknowledgments: The authors are grateful to the Rector of the University of Gdansk, Professor Piotr Stepnowski as well as to the Dean of the Faculty of Economics, Professor Monika Bąk for supporting and sponsoring the work.

Conflicts of Interest: The authors declare no conflict of interest.

Abbreviations

AIS	automatic identification system
CO_2	carbon dioxide
EEDI	Energy Efficiency Design Index
GHG	greenhouse gas
GIS	geographic information systems
HFO	heavy fuel oil
IMO	International Maritime Organization
LNG	liquefied natural gas
LSHFO	low sulphur heavy fuel oil
LSMGO	low sulphur marine gasoil
MAE	Med Atlantic Ecobonus
MCR	maximum continuous rating
MDO	marine diesel oil
MGO	marine gasoil
NO_x	nitrous oxide
PM	particulate matter
SCR	selective catalytic reduction
SECA	Sulphur Emission Control Area
SEEMP	Ship Energy Efficiency Management Plan
SO_x	sulphur oxide
TEU	twenty-foot equivalent unit
VTMS	vessel traffic management system

References

1. International Maritime Organization. Resolution MEPC.304 (72) (adopted on 13 April 2018). In *Initial IMO Strategy on Reduction of GHG Emissions from Ships*; International Maritime Organization: London, UK, 2018.
2. IHS Markit Portal. IHS Online Database. Available online: https://maritime.ihs.com/EntitlementPortal/Home/Index (accessed on 30 June 2019).
3. Johansson, L.; Jalkanen, J.-P.; Kukkonen, J. Global assessment of shipping emissions in 2015 on a high spatial and temporal resolution. *Atmos. Environ.* **2017**, *167*, 403–415. [CrossRef]
4. Maloni, M.; Paul, J.A.; Gligor, D.M. Slow steaming impacts on ocean carriers and shippers. *Marit. Econ. Logist.* **2013**, *15*, 151–171. [CrossRef]
5. Aydin, N.; Lee, H.; Mansouri, S.A. Speed optimization and bunkering in liner shipping in the presence of uncertain service times and time windows at ports. *Eur. J. Oper. Res.* **2017**, *259*, 143–154. [CrossRef]
6. World Health Organization. Air Quality and Health, Fact Sheet N°313, Updated August 2008. Available online: https://www.who.int/mediacentre/factsheets/fs313/en (accessed on 10 November 2019).
7. World Health Organization. Ambient (Outdoor) Air Pollution. 2018. Available online: https://www.who.int/en/news-room/fact-sheets/detail/ambient-(outdoor)-air-quality-and-health (accessed on 29 December 2019).
8. European Environment Agency. Air Pollution by Ozone and Health. 2012. Available online: https://www.eea.europa.eu/data-and-maps/indicators/air-pollution-by-ozone-1/assessment#tab-related-briefings (accessed on 29 December 2019).

9. European Environment Agency. Air Pollution Due to Ozone: Health Impacts and Effects of Climate Change. 2015. Available online: https://www.eea.europa.eu/data-and-maps/indicators/air-pollution-by-ozone-2/assessment (accessed on 10 January 2020).
10. World Health Organization. *Global Health Risks: Mortality and Burden of Disease Attributable to Selected Major Risks*; World Health Organization Press: Geneva, Switzerland, 2009.
11. World Health Organization. *Review of Evidence on Health Aspects of Air Pollution: REVIHAAP Project*; WHO Regional Office for Europe: Copenhagen, Denmark, 2013.
12. Intergovernmental Panel on Climate Change. *AR6 Climate Change 2021: The Physical Science Basis—IPCC*; Intergovernmental Panel on Climate Change: Geneva, Switzerland, 2017.
13. Bouman, E.A.; Lindstad, E.; Rialland, A.I.; Strømman, A.H. State-of-the-art technologies, measures, and potential for reducing GHG emissions from shipping—A review. *Transp. Res. Part D Transp. Environ.* **2017**, *52*, 408–421. [CrossRef]
14. Moreno-Gutiérrez, J.; Pájaro-Velázquez, E.; Amado-Sánchez, Y.; Rodríguez-Moreno, R.; Calderay-Cayetano, F.; Durán-Grados, V. Comparative analysis between different methods for calculating on-board ship's emissions and energy consumption based on operational data. *Sci. Total Environ.* **2019**, *650*, 575–584. [CrossRef]
15. IMO. Sulphur Oxides (SOx) and Particulate Matter—Regulation 14. 2020. Available online: https://www.imo.org/en/OurWork/Environment/PollutionPrevention/AirPollution/Pages/Sulphur-oxides-(SOx)%20Regulation-14.aspx (accessed on 23 January 2020).
16. IMO. Nitrogen Oxides (NOx)—Regulation 13. 2020. Available online: https://www.imo.org/en/OurWork/Environment/PollutionPrevention/AirPollution/Pages/Nitrogen-oxides-(NOx)%20Regulation-13.aspx (accessed on 23 January 2020).
17. Nunes, R.A.O.; Alvim-Ferraz, M.C.M.; Martins, F.G.; Sousa, S.I.V. The activity-based methodology to assess ship emissions—A review. *Environ. Pollut.* **2017**, *231*, 87–103. [CrossRef]
18. Song, S.K.; Shon, Z.H. Current and future emission estimates of exhaust gases and particles from shipping at the largest port in Korea. *Environ. Sci. Pollut. Res.* **2014**, *21*, 6612–6622. [CrossRef]
19. Eyring, V.; Isaksen, I.; Berntsen, T.; Collins, W.; Corbett, J.; Endresen, O.; Grainger, R.; Moldanova, J.; Schlager, H.; Stevenson, D. Transport impacts on atmosphere and climate: Shipping. *Atmos. Environ.* **2009**, *44*, 4735–4771. [CrossRef]
20. Coello, J.; Williams, I.; Hudson, D.A.; Kemp, S. An AIS-based approach to calculate atmospheric emissions from the UK fishing fleet. *Atmos. Environ.* **2015**, *114*, 1–7. [CrossRef]
21. Maragkogianni, A.; Papaefthimiou, S.; Zopounidis, C. *Mitigating Shipping Emissions in European Ports: Social and Environmental Benefits*; Springer: Cham, Switzerland, 2016; pp. 1–9.
22. Corbett, J.J.; Lack, D.A.; Winebrake, J.J.; Harder, S.; Silberman, J.A.; Gold, M. Arctic shipping emissions inventories and future scenarios. *Atmos. Chem. Phys.* **2010**, *10*, 9689–9704. [CrossRef]
23. Endresen, Ø.; Sørgard, E.; Sundet, J.K.; Dalsøren, S.B.; Isaksen, I.S.A.; Berglen, T.F.; Gravir, G. Emission from international sea transportation and environmental impact. *J. Geophys. Res. Atmos.* **2003**, *108*. [CrossRef]
24. Eyring, V.; Köhler, H.W.; Van Aardenne, J.; Lauer, A. Emissions from international shipping: 1. The last 50 years. *J. Geophys. Res. Atmos.* **2005**, *110*. [CrossRef]
25. Psaraftis, H.N.; Kontovas, C.A. CO_2 emission statistics for the world commercial fleet. *WMU J. Marit. Aff.* **2009**, *8*, 1–25. [CrossRef]
26. Cariou, P.; Parola, F.; Notteboom, T. Towards low carbon global supply chains: A multi-trade analysis of CO_2 emission reductions in container shipping. *Int. J. Prod. Econ.* **2019**, *208*, 17–28. [CrossRef]
27. BSR Clean Cargo Working Group (2017). 2016 Global Maritime Trade Lane Emission Factors. Available online: https://www.bsr.org/reports/BSR_CCWG_2016_Global_Maritime_Trade_Lane_Emissions_Factors.pdf (accessed on 10 January 2020).
28. Corbett, J.J.; Köhler, H.W. Updated emissions from ocean shipping. *J. Geophys. Res.* **2003**, *108*. [CrossRef]
29. Liu, H.; Meng, Z.H.; Shang, Y.; Lv, Z.F.; Jin, X.X.; Fu, M.L.; He, K.B. Shipping emission forecasts and cost-benefit analysis of China ports and key regions' control. *Environ. Pollut.* **2018**, *236*, 49–59. [CrossRef]
30. Liu, J.; Duru, O. Bayesian probabilistic forecasting for ship emissions. *Atmos. Environ.* **2020**, *231*, 117540. [CrossRef]
31. IMO. UN Agency Pushes Forward on Shipping Emissions Reduction. Available online: https://www.imo.org/en/MediaCentre/PressBriefings/Pages/11-MEPC-74-GHG.aspx (accessed on 10 January 2020).
32. Comer, B.; Olmer, N.; Mao, X.; Roy, B.; Rutherford, D. *Black Carbon Emissions and Fuel Use in Global Shipping 2015*; International Council on Clean Transportation: Washington, DC, USA, 2017.
33. UNCTAD/RMT/2018. UNCTAD—Review of Maritime Transport 2018. Available online: https://unctad.org/en/PublicationsLibrary/ (accessed on 5 December 2019).
34. Rouse, H.; Koloseus, H.J.; Davidian, J. The Role of The Froude Number In Open-Channel Resistance. *J. Hydraul. Res.* **1963**, *1*, 14–19. [CrossRef]
35. Mayer, F.T.; Fringer, O.B. An unambiguous definition of the Froude number for lee waves in the deep ocean. *J. Fluid Mech.* **2017**, *831*, R3. [CrossRef]
36. Hager, W.H.; Castro-Orgaz, O. William Froude and the Froude Number. *J. Hydraul. Eng.* **2017**, *143*, 02516005. [CrossRef]
37. Med Atlantic Ecobonus Project. Puertos del Estado, Spain. 2018. Available online: http://mae-project.eu/ (accessed on 10 November 2019).
38. Azzara, A.; Minjares, R.; Rutherford, D.; Needs and Opportunities to Reduce Black Carbon Emissions from Maritime Shipping. ICCT Working Paper 2015-2. Available online: https://theicct.org/publications/needs-and-opportunities-reduce-black-carbon-emissions-maritime-shipping (accessed on 24 January 2020).

39. Fuglestvedt, J.S.; Shine, K.P.; Berntsen, T.; Cook, J.; Lee, D.S.; Stenke, A.; Skeie, R.B.; Velders, G.J.M.; Waitz, I.A. Transport impacts on atmosphere and climate: Metrics. *Atmos. Environ.* **2010**, *44*, 4648–4677. [CrossRef]
40. Reisinger, A.; Meinshausen, M.; Manning, M.; Bodeker, G. Uncertainties of global warming metrics: CO_2 and CH_4. *Geophys. Res. Lett.* **2010**, *37*. [CrossRef]
41. Marine Traffic Website. Marine Traffic: Global Ship Tracking Intelligence, AIS Marine Traffic. Available online: https://www.marinetraffic.com/en/ais/home/centerx:-12.0/centery:25.0/zoom:4 (accessed on 10 December 2018).
42. Winther, M.; Christensen, J.H.; Plejdrup, M.S.; Ravn, E.S.; Eriksson, Ó.F.; Kristensen, H.O. Emission inventories for ships in the arctic based on satellite sampled AIS data. *Atmos. Environ.* **2014**, *91*, 1–14. [CrossRef]
43. Kumar, S. Acid Rain-The Major Cause of Pollution: Its Causes, Effects. *Int. J. Appl. Chem.* **2017**, *13*, 53–58.
44. Schrooten, L.; De Vlieger, I.; Panis, L.I.; Chiffi, C.; Pastori, E. Emissions of maritime transport: A European reference system. *Sci. Total Environ.* **2009**, *408*, 318–323. [CrossRef]
45. Welaya, Y.M.A.; El Gohary, M.M.; Ammar, N.R. A comparison between fuel cells and other alternatives for marine electric power generation. *Int. J. Nav. Archit. Ocean Eng.* **2011**, *3*, 141–149. [CrossRef]
46. Hordeski, M.F. *Alternative Fuels: The Future of Hydrogen*; Fairmont Press: Lilburn, GA, USA, 2008.
47. Blanco, H.; Nijs, W.; Ruf, J.; Faaij, A. Potential for hydrogen and Power-to-Liquid in a low-carbon EU energy system using cost optimization. *Appl. Energy* **2018**, *232*, 617–639. [CrossRef]
48. Marchenko, O.V.; Solomin, S.V. The future energy: Hydrogen versus electricity. *Int. J. Hydrog. Energy* **2015**, *40*, 3801–3805. [CrossRef]
49. Staffell, I.; Scamman, D.; Velazquez Abad, A.; Balcombe, P.; Dodds, P.E.; Ekins, P.; Shahd, N.; Warda, K.R. The role of hydrogen and fuel cells in the global energy system. *Energy Environ. Sci.* **2019**, *12*, 463–491. [CrossRef]
50. Hansson, J.; Månsson, S.; Brynolf, S.; Grahn, M. Alternative marine fuels: Prospects based on multi-criteria decision analysis involving Swedish stakeholders. *Biomass Bioenergy* **2019**, *126*, 159–173. [CrossRef]
51. Smith, T.; Day, S.; Bucknall, R.; Mangan, J.; Dinwoodie, J.; Landamore, M.; Turan, O. *Low Carbon Shipping: A Systems Approach*; Safety4Sea Press: Jakarta, Indonesia, 2014.
52. United Nations Resolution A/RES/70/1 25 September 2015. United Nations Sustainable Development Goals. Available online: https://www.un.org/ga/search/view_doc.asp?symbol=A/RES/70/1&Lang=E (accessed on 3 May 2019).
53. Levinson, M. *The Box: How the Shipping Container Made the World Smaller and the World Economy Bigger*; Princeton University Press: Princeton, NJ, USA, 2016.

 *energies*

Article

Effects of Regional Innovation Capability on the Green Technology Efficiency of China's Manufacturing Industry: Evidence from Listed Companies

Yu Fu [1,2], Agus Supriyadi [1,2,3] , Tao Wang [1,2,*], Luwei Wang [1,2] and Giuseppe T. Cirella [4]

[1]   School of Geography, Nanjing Normal University, Nanjing 210023, China; 181301005@stu.njnu.edu.cn (Y.F.); 31173018@stu.njnu.edu.cn (A.S.); 191301012@stu.njnu.edu.cn (L.W.)

[2]   Jiangsu Center for Collaborative Innovation in Geographical Information Resource Development and Application, Nanjing 210023, China

[3]   Inspectorate Banjar City Government, Banjar West Java 46311, Indonesia

[4]   Faculty of Economics, University of Gdansk, 81-824 Sopot, Poland; gt.cirella@ug.edu.pl

*   Correspondence: wangtao@njnu.edu.cn; Tel.: +86-25-8589-1745

Received: 20 August 2020; Accepted: 15 October 2020; Published: 19 October 2020

Abstract: The purpose of the "Made in China 2025" strategy is to enhance the innovation capabilities of the local manufacturing industry and achieve green and sustainable development. The role of innovation in the development of manufacturing is a hotspot in academic research, though only a few studies have analyzed the interaction between green technology manufacturing efficiency and its external innovation capabilities. This study used the 2011–2017 Chinese A-share listed manufacturing companies as samples to discuss whether regional innovation capabilities can promote the improvement of green technology manufacturing efficiency. The results showed that a significant spatial correlation between regional innovation capability and green technology manufacturing efficiency was prevalent within spatial heterogeneous bounds. In addition, regional innovation capability directly promoted the effective manufacturing of green technology efficiency, which was strongest in the eastern region of the country. Regional innovation capabilities also had a positive effect on human capital and government revenue, thereby further enhancing the green technology efficiency of manufacturing through the intermediary effect. Based on the above conclusions, some policy recommendations are put forward to facilitate the improvement of China's regional innovation capabilities in terms of green technology efficiency in manufacturing.

Keywords: innovation capability; spatial association; industry upgrade; Tobit model; intermediary effect; sustainable development; China

---

1. Introduction

Over the past few decades, many countries have begun to take sustainable action in anticipation of environmental backlash [1]. Global climate change is the basis for a universal awareness of environmental protection for future generations. It also encourages innovation in various facets of human life. Previous research from India highlighted that changes within the energy industry strongly interlinked energy, population, and urbanization [2]. In terms of reducing environmental impacts, the emergence of "green innovation" has increasingly become a popular trend in research, as well as a point of discussion for academics, industry partners, and politicians alike [3,4]. Schiederig et al. [5] summarized green innovation into six essential elements: innovation object (i.e., product, process, service, and method), market orientation, innovation environment, its full life cycle (i.e., as the central consideration for material flow reduction), intention toward the reduction of economic or ecological demands, and green standards (i.e., in terms of the firm). Despite this, the implementation of green innovation often encounters challenges when applied in non-green industries due to the limitation of new resources and the competencies and capabilities of the industry in terms of changes to the production processes [6]. In the application of green innovation, modifying many factors, as well as the role of participants, should be the first step in terms of developing an environmental protection-based system. Reciprocally, the use of the sustainable resources and green technological processes of the system should be highly dependent (i.e., correlative) to the product market [6,7]. As such, green technology plays a vital role in achieving green sustainable development programs.

The National Bureau of Statistics of China announced that China's total manufacturing volume has been ranked first in the world for many consecutive years, in which the added value from the manufacturing industry accounted for 27.17% of GDP in 2019. China's economic growth is greatly influenced by developments in the manufacturing industry. However, its manufacturing industry has a weak independent innovation capacity and a low resource utilization rate, causing a series of environmental pollution-related problems. As a result, the Chinese government has been focusing on formulating policies to increase research and development investment specific to this field [8,9]. In terms of reducing greenhouse gas emissions, as well as pushing for renewable energy, China has begun to develop renewable energy sectors more actively since 2005. This was marked by the issuance of supporting measures and regulations for stimulating renewable energy development [10,11]. Apart from the many factors that motivated the formulation of policies in terms of stimulating these sectors, increases in the budgetary allotment have been extraordinary and have become essential in supporting the effort to save the environment. Innovations in the field of green technology have also led to more effective, efficient, and economical systems [12]. Thus, conducting research on the manufacturing industry's green technology efficiency in correlation with innovation in China, including green technology and its application to support the sustainability of green development programs, is crucial.

The application of green technology in China has been a growing concern for the Chinese government, which has been working on encouraging many enterprises to apply this technology to activities in all regions [13]. With the widespread application of green technology throughout many industries nationwide, a key issue is how to formulate the development of green technology efficiency and measure regional differences in terms of implementation. It becomes apparent that the spatial correlation between the developing regional innovation and the efficiency of applied green technology overlaps. Previous studies have discussed enterprise preparedness for green innovation in terms of technological, environmental, organizational, and policy constraints [11,13,14]. However, limited studies have been conducted to reveal the regional innovation related to green technology efficiency and its implementation. First, most of the studies on the relationship between innovation and technological efficiency are from the perspective of enterprise innovation investment, with limited studies from a research perspective. Second, the evaluation indicators of existing regional innovations are expressed mostly in terms of the number of published papers or patent authorizations. The indicator system is relatively unrepresentative, which makes it difficult to reflect upon the actual (i.e., complete) regional innovation capability. Finally, the role of geographic space is generally ignored, and the regional heterogeneity and spatial correlation between research elements are not considered. Listed companies, as representatives of high-quality enterprises, need to accept external auditing, which involves disclosing its information publicly such that its data is transparent and easily accessible.

This study used the spatial autocorrelation and Tobit model to select 2011–2017 Chinese A-share listed manufacturing companies as the research sample to answer whether regional innovation ability directly affects the green technology efficiency of China's manufacturing industry. In addition, regional innovation has a significant spillover effect that inevitably affects factors such as human capital, government revenue, and the waste treatment rate [15]. Can changes in the above factors indirectly affect the green technology efficiency of manufacturing? Differences in resource endowments in various regions will cause differences in innovation capability and technical efficiency [16]. As such, it is necessary to explore the heterogeneity of the impact of different regions' innovation capabilities on their green technology manufacturing efficiency. Thus, the foregoing is an attempt to analyze the impact of regional innovation capabilities on the efficiency of green technology in China's manufacturing industry, where three aspects are examined: (1) direct effects, (2) spatial heterogeneity, and (3) indirect effects. The specific research framework is shown in Figure 1.

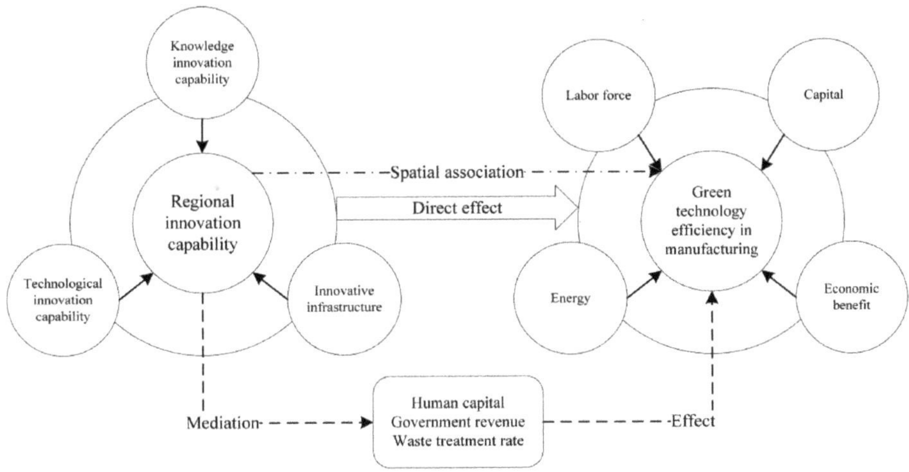

Figure 1. Research framework.

The contributive factors of this research were central to the following three viewpoints. First, it combined geographical and economic methods to comprehensively explore the interaction between regional innovation capabilities and green technology manufacturing efficiency, as well as compensated for the shortcomings of the existing literature. Second, it innovatively explored the effects of manufacturing external innovation capability on its green technology efficiency, broke the constraints of traditional enterprise perspective research, focused on urban units, and expanded the research perspective of the interaction between innovation and technological efficiency. Third, it sorted out the direct and intermediate roles between regional innovation capabilities and green technology manufacturing efficiency by providing theoretical and practical guidance for enterprises and government, which can help with decision-making and efficiency. The paper is organized as follows: Section 2 reviews the literature, Section 3 introduces the indicator construction and data sources, Section 4 lists the research methods, Section 5 illustrates the spatial correlation and empirical analysis results regarding regional innovation capability and green technology efficiency of China's manufacturing industry, and Section 6 elucidates the concluding remarks and policy implications.

2. Literature Review and Theoretical Framework

2.1. Literature Review

The significant impact of climate change has brought forth a worldwide paradigm shift, especially in terms of the use of green technology. Previous research has shown that a government's essential role in encouraging the application of green technology in industry varies greatly [17,18]. The application of green technology will differ and be influenced by the capacity and ability of institutions that are specific to each country. Green technology policy requires not only consistency from the industry but also political stability from political actors in a country. The framework of its development plan needs to be sustainable, as well as flexible, and should not be changed due to political intervention [17].

Therefore, political, economic, and sustainable promotion in a country should encourage the industry and society at large to create more effective and efficient innovations in the field of green technology.

Since 2006, when the Chinese government began to adopt strategic policies, by encouraging innovation-driven development, it began driving significant social and economic development by focusing on industries such that they could develop rapidly through extraordinary innovation [19,20]. Government funding and tax incentives played, and continue to play, an important role in promoting green technology innovation. This interaction, i.e., between government regulations and government research, has been a vital part of promoting green product innovation [18]. Various studies have indicated that government regulations contribute significantly to low-carbon technology innovation compared with technology pushes and the market [21]. Based on the evidence, the importance of government regulations and initiatives to promote innovation (i.e., especially in terms of green technology applications) in industry can be observed.

According to the results of a study conducted in a democratic country, no direct correlation between democracy and innovation could be found [22]. The national innovation activities are located mostly in areas where companies, universities, and governments interact directly and the encouragement of the local government in such innovation is crucial [23]. Thus, understanding how the influence of innovation in certain regions effectively and efficiently supports and maintains the sustainable implementation of green technology becomes paramount to its realization and advancement [22,23].

In terms of China, the manufacturing industry has not only become its main economic contributor but also the most important contributor to energy consumption and environmental pollution, which has signaled an urgency for the development of green technologies in manufacturing [24]. Previous studies showed that the development of green growth was influenced considerably by a paradigm shift in China's manufacturing industry, which had begun to move toward a green-oriented focus regarding innovation. At the same time, however, the disparity in the inefficiencies between the western and eastern regions of China was also increasing [8,25]. Different research concluded that green growth performance in China is profoundly influenced by regional innovation capacity, technological innovation, and the technical and institutional instrument coordination [26]. Based on this evidence, regional innovation is also key to the success of green technology implementation. However, differing results experienced in each region due to varying causes, e.g., geographical location, economic development, flexible innovation mechanisms, the role of a strong market economy, intellectual property protection, corporate credit, market financing, and government public education investment, should be taken into account [26,27].

Strict environmental regulations and openness to external international policies have encouraged, and to some degree compelled, the application of green technology in China's manufacturing industry. This increase in green development, especially in China's high-end manufacturing industry, is not in line with the innovation and efficiency of green technology compared with the traditional manufacturing industry [28]. This continual adjustment has resulted in a win–win scenario for its manufacturing industry. Other research from Shenzhen showed that regional innovation is significantly dominated by large, high-tech firms and the dynamics of high-tech entrepreneurship, which is closely related with high impact human capital (i.e., high-quality talents) [29]. In summary, the topic of green development, which includes innovation and green technology and development, has attracted the attention of many scholars, industry, and the government. The growing awareness and importance within the field exemplify key issues in terms of environmental and human sustainability for future generations. Further research recommendations include conducting more in-depth, innovation-oriented, region-based studies and exploring the effects on the efficiency of green technology, as well as how the effects of intermediary variables and many others have been echoed [29,30]. Based on a review of the existing literature, this research sought to contribute to the knowledge base and fill the literature gap with a focused study on manufacturing industries in China. This is important because this industry is not only a significant contributor to the economic growth of China but also a significant contributor to pollution and related environmental issues.

2.2. Theoretical Framework

2.2.1. Direct Effect

Regional innovation is the fundamental driving force for promoting social and economic development. It started with investments in technology development, with the ultimate goal of converting technological outputs into commercial value [31]. The direct impact of regional innovation capability on the efficiency of manufacturing green technology is mainly manifested in (1) enterprises innovating energy-saving and environmental protection technologies that change production methods and eliminating high-energy-consuming old technologies through new technologies, and (2) reducing energy consumption and pollutant emissions in the production process and realizing the improvement of green technology efficiency in manufacturing [32]. Based on these direct effects, this study proposed:

Hypothesis 1. *Regional innovation capabilities can directly promote the progress of green technology manufacturing efficiency.*

2.2.2. Spatial Heterogeneity

The location conditions of different spatial regions are heterogeneous because different locations have different attributes or qualifications and the elements provided for human production are also different [33]. The geographical differences in the manufacturing industry will lead to differences in the number of universities and scientific research institutes, traffic accessibility, government technology policies, infrastructure, and other innovation factors, which will affect the differences in the absorption, creation, and transformation of green technology efficiency in different regions. Therefore, in terms of spatial heterogeneity, this study proposed:

Hypothesis 2. *The impact of different regional innovation capabilities on the efficiency of manufacturing green technology has spatial heterogeneity.*

2.2.3. Indirect Effect

Regional innovation capabilities have an important impact on human capital, government revenue, and waste treatment rates, and these factors inevitably have an indirect spillover effect on green technology manufacturing efficiency. First, regions with a higher level of regional innovation capability can attract more high-quality talents who bring advanced production technology and management experience to promote the improvement of green technology efficiency in manufacturing [34]. Second, the improvement of regional innovation capability adds vitality to social and economic development, thereby promoting the increase of government revenue, and providing strong support and guarantees for the update of green technology manufacturing efficiency. Third, the innovation of resource recycling and waste treatment technology can improve the reuse and distribution of resources, increase the efficiency of raw materials and energy use, and thus promote the progress of green technology efficiency in manufacturing [35]. As a result of these indirect effects, this study proposed:

Hypothesis 3. *Regional innovation capabilities can indirectly promote the progress of green technology efficiency in manufacturing through human capital, government revenue, and the waste treatment rate.*

3. Indicator Construction and Data Sources

3.1. Construction of a Green Technology Efficiency Index System for China's Manufacturing Industry

The green technology efficiency of the manufacturing industry is essential for obtaining the largest possible economic output (i.e., in terms of manufacturing) with the least input from production factors and the lowest environmental cost, thereby reflecting the relationship between the manufacturing economy and resource utilization and environmental protection [36]. Jorgenson [37] decompose inputs into three major elements: labor force, capital, and energy. This study took the total number of employees, fixed asset net worth, and hydroelectric energy consumption of the listed companies in the manufacturing industry as input indicators, as well as the operating income as the measure of economic benefit output, to build the green technology efficiency index system of Chinese manufacturing, based on input–output theory. The methods of measuring industrial technical efficiency were divided into data envelopment analysis and stochastic frontier analysis. Because traditional data envelopment analysis cannot process time series data, it is not suitable for the study of efficiency-influencing factors using panel data and ignores the effects of random factors on the efficiency estimation; contrarywise, the stochastic frontier analysis method compensates for the above shortcomings and is employed as a viable measuring approach.

3.2. Construction of the Regional Innovation Capability Index System

Regional innovation capability is aimed at enhancing the driving force of regional economic growth, giving full play to the enthusiasm of regional innovation behaviors, efficiently allocating regional innovation resources, and transforming innovative ideas into the comprehensive capabilities of new technologies, new products, new processes, and new services [38]. This study referred to the index systems of the OECD Innovation Index and the China Science and Technology Development Strategy Research Group. Based on the principles of scientificity and representativeness, this research comprehensively built a regional innovation capability index system from the three dimensions of knowledge production capability, technological innovation capability, and innovative infrastructure, with a total of 12 indicators (Table 1). Then, the entropy method was used to calculate the comprehensive index of each city's innovation capability (*INN*).

Table 1. Index system for innovation capacity.

Target Layer	Feature Layer	Indicator Layer	Unit
Regional innovation capability	Knowledge innovation capability	Number of ordinary colleges and universities	Institute instead
		Number of full-time teachers in ordinary colleges and universities	Person
		Public library book collection	Ten-thousand volumes
		Proportion of education expenditure that made up local public financial expenditure	%
	Technological innovation capability	Number of employees in scientific research, technical services, and geological exploration	Person
		The proportion of foreign capital actually used in that year as a percentage of GDP	%
		Total patent grants at the end of the year	Grant
		Science and technology expenditure accounts for the local public finance expenditure	%
	Innovative infrastructure	Greening rate of the built-up area	%
		Telecommunications revenue	Ten-thousand yuan
		Number of Internet broadband users	Ten-thousand households
		Urban road area at the end of the year	Ten-thousand square meters

3.3. Control Variable Setting

Controlling other variables that affect the efficiency of green technology in the manufacturing industry is necessary to obtain unbiased estimation results. This research systematically and comprehensively selected control variables from the perspective of manufacturing (i.e., city) and itself (i.e., industry). First, the economic development level in the form of regional gross domestic product (*GDP*) and foreign cooperation (*FC*) indicators were selected at the city level. Economic development provides abundant capital support for the rise of green technology efficiency, and regions with high levels of economic development having higher levels of industrialization and relatively complete environmental policies. Hence, regional *GDP* was used to measure the level of urban economic development [39]. The cooperation between local and foreign enterprises, and as a result, the knowledge and technology spillover effect in mutual exchange may promote the improvement of production efficiency. However, when foreign capital enters, it will also squeeze the market share of domestic enterprises, thereby inhibiting their development, which is not conducive to the improvement of the local green technology manufacturing efficiency. The number of foreign direct investment contract projects was used to measure the variables of foreign cooperation. Second, government subsidies (*GI*) and manufacturing scale (*MS*) indicators were selected at the industrial level. Government grants to enterprises reduce the cost and risk of upgrading the technology efficiency of enterprises but they may also have a "crowding out effect" [40]. The level of government subsidies to manufacturing is measured by government investment in enterprises. If a certain scale of the manufacturing industry is formed in the city, it will inevitably bring about the concentration of enterprises in the upstream and downstream industrial chains and the improvement of the infrastructure level, which will contribute to the improvement of the technological efficiency of those enterprises [28]. On the downside, the continuous expansion of the manufacturing industry in cities may bring about environmental pollution, vicious competition, and reduction of market management and control capabilities, which will damage the efficiency of green technology. As a result, the value of the total assets of manufacturing enterprises was employed to represent the manufacturing scale variables.

3.4. Selection of Intermediary Variables

Human capital, government revenue, and the waste treatment rate as intermediary variables were used to explore the indirect impact mechanisms of regional innovation capability on green technology manufacturing efficiency. Human capital (*HC*) was measured using the number of students in ordinary colleges and universities. Innovation activities require a large number of high-quality, professional talent, and as such, the improvement of human capital will promote the upgrading of green technology manufacturing efficiency [35]. Government revenue (*GR*) was characterized by public fiscal revenue. Schumpeter's innovation theory states that innovation is the endogenous driving force for economic development. The improvement of the economic development level will inevitably bring about an increase in government revenue. The government builds a complete green ecology for the manufacturing industry through fiscal and tax policies [39]. The waste treatment rate (*WTR*) variable was obtained using the entropy method to comprehensively calculate the three indicators of general industrial solid waste: comprehensive utilization rate, centralized treatment rate of the sewage treatment plant, and harmless treatment rate of domestic garbage. Innovation promotes the technological upgrading of the waste treatment industry, while the technological innovation of the waste treatment rate promotes the improvement of the overall green technology efficiency of the manufacturing industry.

3.5. Data Sources

This study merged listed manufacturing companies in Chinese prefecture-level cities and above as research units. The Renminbi common stocks (i.e., A-shares) only include companies registered in mainland China, Chinese Taiwan, Hong Kong, and Macau were temporarily not listed as analysis objects. The observation period of this study was 2011–2017 to address the issues on the availability and completeness of data since Chinese listed companies had fewer requirements regarding the disclosure of environmental information before 2011. The data of listed companies in China's manufacturing industry were obtained from the Choice database (i.e., http://choice.eastmoney.com/), and the total amount of urban patents obtained was from the Chinese Research Data Service Platform (i.e., https://www.cnrds.com/Home/Index/). All other data were derived from the "China City Statistical Yearbook" of the corresponding year. The missing data were extrapolated according to the interpolation method. The vector data of national boundaries and territorial boundaries were derived from the national 1:1 million basic geographic databases published by the China National Basic Geographic Information Center in 2017 (i.e., http://www.webmap.cn/).

4. Research Methods

4.1. Entropy Method

The entropy method determines the weight of an index based on the influence of the relative change degree of the index on the overall system. It is an objective weighting method that comprehensively considers various factors and is suitable for multiple indicators. The entropy method has certain objectivity and scientific characteristics because it can overcome randomness problems in terms of subjective weighting and solve the problem of information overlap between multiple index variables. Taking this into account, this study constructed a regional innovation capability index system, standardized the original data, and used the entropy method to calculate the innovation capability of each city. The calculation steps of the entropy method are as follows.

First, calculate the proportion of the jth index in city i (i.e., Equation (1)); second, calculate the entropy value of the indicator (i.e., Equation (2)); third, calculate the information utility value of the jth indicator (i.e., Equation (3)); fourth, calculate the weight of the index x_i (i.e., Equation (4)); finally, calculate the level of innovation capability of each city (i.e., Equation (5)).

$$P_{ij} = x_{ij} / \sum_{i=1}^{n} x_{ij}, \tag{1}$$

$$e_j = -k \sum_{i=1}^{n} P_{ij} \ln P_{ij}, \ k = 1/\ln(n), \tag{2}$$

$$g_j = 1 - e_j, \tag{3}$$

$$w_j = g_i / \sum_{j=1}^{p} g_j, \tag{4}$$

$$INN = \sum_{j=1}^{p} w_j x_{ij}. \tag{5}$$

4.2. Stochastic Frontier Analysis

The stochastic frontier production function model was originally proposed by Aigner et al. [41]. It is a typical parameter efficiency measurement method that can distinguish production function errors into two parts: random errors caused by uncontrollable random factors and management errors caused by technical inefficiency. Battese and Coelli [42] introduced panel data and proposed that the BC model of stochastic frontier analysis be used to measure the industrial technical efficiency, resource utilization efficiency, and urban development efficiency. This study drew on the research method of Yang et al. [43] to construct a stochastic frontier production function to measure the green technology efficiency of China's manufacturing industry. The formulation, i.e., Equations (6)–(10), was as follows:

$$\ln(Y_{it}) = \beta_0 + \beta_1 \times (X_{1it}) + \beta_2 \times \ln(X_{2it}) + ... + \beta_n \times (X_{nit}) + v_{it} - u_{it}, \tag{6}$$

$$TE_{it} = \exp(-u_{it}), \tag{7}$$

$$\beta(t) = \exp[-\eta \times (t - T)], \tag{8}$$

$$u_{it} = \beta(t) \times u_i, \tag{9}$$

$$\gamma = \frac{\sigma_u^2}{\sigma_v^2 + \sigma_u^2}, \gamma \in [0,1], \tag{10}$$

where Y is the output variable; X is an input variable; i is a research unit; t is time; T represents the time trend of technological changes; β_0 is the intercept term; $\beta_1, \beta_2, ...\beta_n, \eta$, and γ denote the parameters to be estimated; v_{it} is a random error term subject to a normal distribution $N(0, \sigma_v^2)$; u_{it} denotes a non-negative management error term, subject to a non-negative truncated normal distribution $N(0, \sigma_u^2)$; TE indicates the technical efficiency value.

4.3. Spatial Autocorrelation Method

Spatial dependence means that a certain element does not exist independently in a spatial unit but is instead related to the adjacent spatial unit, where spatial autocorrelation is a quantitative index to check the strength of the association between the attribute value of a certain element and the attribute value of the adjacent space [44].

4.3.1. Local Indicators of Spatial Association (LISA)

Local spatial autocorrelation is used to measure the degree of similarity (i.e., positive correlation) or difference (i.e., negative correlation) of a certain attribute at its local position with its neighboring spatial unit attributes. The heterogeneous characteristics of spatial elements can be understood more fully, including how spatial dependence changes with location [45]. The use of local spatial autocorrelation methods to explore the spatial distribution characteristics of the regional innovation capability and the green technology efficiency of Chinese manufacturing was applied. Local Moran's I is a quantitative index of local spatial autocorrelation; its calculation is formulated using Equation (11):

$$I_i = \frac{(x_i - \bar{x})}{S^2} \sum_{j \neq i}^{n} w_{ij}(x_j - \bar{x}), \bar{x} = \frac{1}{n} \sum_{i=1}^{n} x_i, S^2 = \frac{1}{n} \sum_{i} (x_i - \bar{x})^2, \tag{11}$$

where n indicates the total number of observation points; w_{ij} is the space weight; x_i and x_j denote the spatial elements i and j of the attribute, respectively; $\bar{x}$ represents the average value of the attribute; S^2 is the variance of the attribute.

4.3.2. Bivariate Moran's I

A bivariate global spatial autocorrelation can be used to show the overall agglomeration characteristics and spatial correlation between two attributes. The expression can be found in Equation (12):

$$I = \sum_{i=1}^{n} \sum_{j=1}^{n} w_{ij}(x_i - \bar{x})(y_j - \bar{y}) / S^2 \sum_{i=1}^{n} \sum_{j=1}^{n} w_{ij}, \tag{12}$$

where I is a bivariate global spatial autocorrelation coefficient that represents the correlation between the overall spatial distribution of the independent variable of spatial element i and the dependent variable of the spatial element j, x_i and y_j are the spatial elements i and j of the attribute, and other symbols are the same as in Equation (11) above [46].

4.3.3. Bivariate Local Moran's I

A bivariate global spatial autocorrelation can be used to show the overall agglomeration characteristics and spatial correlation between two attributes. Equation (13) was used, as follows:

$$I_{i,j} = z_i \sum_{j=1}^{n} w_{ij} z_j, \tag{13}$$

where z_i and z_j represent the standardized variance values of the spatial elements i and j, respectively, and $I_{i,j}$ is the local correlation between the independent variable of region i and the dependent variable of region j, according to which, the spatial elements can be divided into the four types of agglomeration, namely, H–H (high–high), L–L (low–low), H–L (high–low), and L–H (low–high). The H–H cluster means the independent variable value of spatial element i and the dependent variable value of spatial element j are both high, the L–L cluster means that both attribute values are low, where the H–H and L–L clusters mean that the independent variable of space element i has a positive effect on the dependent variable of spatial element j [47].

4.4. Empirical Methods

In this study, the explained variable of China's green technology efficiency in manufacturing had an obvious range limit. This value was represented in the range of zero to one. The results of the traditional ordinary least squares method may be biased. Therefore, our benchmark regression used Tobit estimation, a common processing method for censored data, and considered the upper and lower thresholds of the data. Based on the aforementioned literature, the following direct impact measurement model was constructed:

$$Te_{it} = \beta_0 + \beta_1 Ia_{it} + \beta_n Control_{it} + \varepsilon_{it}, \tag{14}$$

where Te_{it} represents city i's green technology manufacturing efficiency level during period t, Ia_{it} measures the innovation capability index of the city i in period t, β_0 is the intercept term, β_1 denotes the regression coefficient of the innovation ability for the green technology manufacturing efficiency, $Control_{it}$ is the set of control variables, and ε_{it} is a random disturbance.

This study added intermediary variables *Med* to construct the following intermediary effect measurement model to further explore the indirect effect of the regional innovation capability on the green technology efficiency of manufacturing, as formulated in Equations (15) and (16):

$$Med_{it} = \beta_0 + \beta_1 Ia_{it} + \varepsilon_{it}, \tag{15}$$

$$Te_{it} = \beta_0 + \beta_1 Ia_{it} + \beta_2 Ts_{it} + \varepsilon_{it}. \tag{16}$$

Some unobservable missing variables or a two-way causal relationship between innovation capabilities and green technology manufacturing efficiency may lead to endogenous problems. This study referenced Lin and Tan [48] to add terrain slope tool variables Ts and used a panel data two-stage least squares method to solve possible endogenous problems. The model settings, i.e., Equation (17), were applied as follows:

$$Te_{it} = \beta_0 + \beta_1 Ia_{it} + \beta_2 Ts_{it} + \beta_n Control_{it} + \varepsilon_{it}. \tag{17}$$

5. Research Results

5.1. Spatial Association Pattern

5.1.1. Spatial Correlation Distribution Characteristics of Green Technology Efficiency and Innovation Capability in the Manufacturing Industry

First, the applicability of the method was checked. The global spatial autocorrelation Moran's I indices of China's green technology manufacturing efficiency and the urban innovation capacity were calculated from 2011 to 2017 with the help of ArcGIS 10.2 software (Environmental Systems Research Institute, Redlands, CA, USA). The results showed that the Moran's I estimates passed the significance level test with a confidence level of 99%, indicating that China's green technology manufacturing efficiency and urban innovation capacity were spatially clustered. Thus, local spatial autocorrelation analysis could be used. This study selected 2011 and 2017 as the beginning and end time nodes, respectively, and drew the LISA cluster map of China's manufacturing industry's green technology efficiency, as well as its regional innovation capability level in terms of the spatial autocorrelation (Figures 2 and 3). This study divided the local spatial units into four types: high–high cluster (i.e., H–H), high–low cluster (i.e., H–L), low–high cluster (i.e., L–H), and low–low cluster (i.e., L–L) to reveal their degree of association and distribution in the local spatial position pattern.

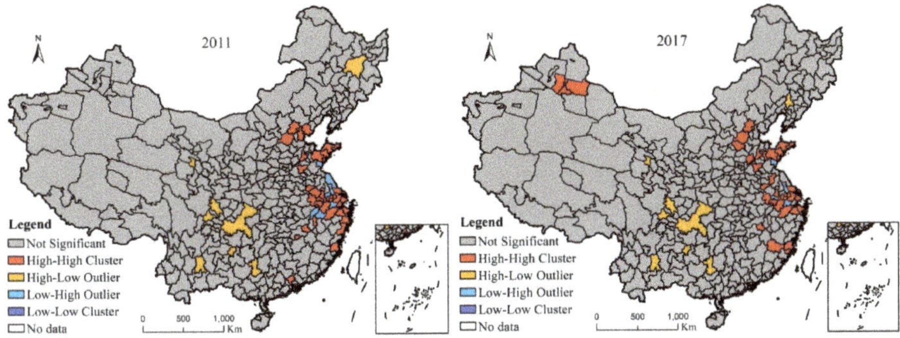

Figure 2. Local indicators of spatial association (LISA) pattern of green technology efficiency in China's manufacturing industry.

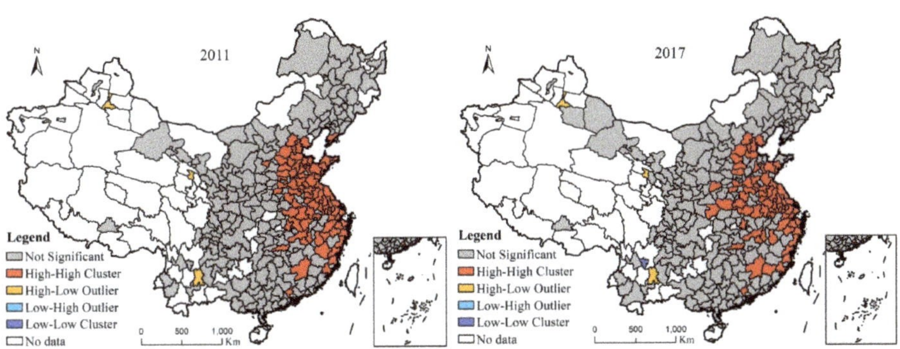

Figure 3. LISA pattern of regional innovation capability.

In general, the local correlation pattern of the green technology efficiency in China's manufacturing industry was characterized by "a large agglomeration and a small dispersion." The H–H cluster areas were located in some cities along the eastern coast of Beijing-Tianjin-Hebei, Shandong Peninsula, Yangtze River Delta, Fujian Province, and Guangdong Province. Such regions had a high level of economic development; large industrial scale; abundant capital; rich, high-quality talent; top infrastructure; advanced and sophisticated industrial technology; a large output of economic benefits, and thus, the green technology manufacturing efficiency levels were in the lead position. The H–H cluster areas in the Pearl River Delta were gradually decreasing, while the number of high and high-concentration areas in inland Xinjiang increased. On the one hand, because of the overall "large input and low output" phenomenon in the Pearl River Delta, the input of production factors, such as labor, land rent, water supply, and electricity, increased such that the traditional manufacturing industry moved out in large numbers, the overall output scale declined, and high-tech industries are still to be developed; as a result, various factors weakened the green technology efficiency in the manufacturing industry throughout the Pearl River Delta region. On the other hand, as the core area of the "Belt and Road" strategy, Xinjiang attracted a large number of manufacturing industries with its preferential policies, low labor and land rents, and abundant resources, which led to the formation of the H–H cluster area of manufacturing industrial green technology efficiency in Urumqi and its surrounding cities.

The L–H cluster areas were attached to the surrounding areas of H–H cluster areas, such as Yancheng, Zhenjiang, Taizhou, Huzhou, Zhoushan, Huangshan, Chizhou, Rizhao, and Ma'anshan. Such cities had relatively poor industrial economic benefits, an irrational industrial structure, and a weak "diffusion effect" from the surrounding developed cities, resulting in a large gap between the green technology efficiency value of neighboring H–H cluster cities, and thus, a "collapsed" formation in terms of space. However, the number of L–H cluster areas gradually decreased, indicating that the technology spillover effect of surrounding cities increased over time and the regional "unbalanced" problem has been alleviated to a certain extent. The H–L cluster areas were scattered in the provincial capital of central cities in the west and northeast, including Chongqing, Chengdu, Guiyang, Kunming, Xining, Harbin, and Shenyang. As an administrative center in less developed regions, the efficiency of green technology is closer to the frontier of production than other surrounding cities. With the existence of the "siphon effect," promoting the improvement of green technology efficiency in the surrounding cities is difficult and the "Matthew effect" eventually appears in the region. During the study period, the spatial units of the H–L cluster did not change and the L–L cluster areas did not appear.

The regional spatial agglomeration characteristics of China's regional innovation capability were remarkable and the H–H cluster was the main distribution type. In 2011, H–H cluster areas were distributed in a "planar form" in Beijing, Tianjin, Hebei, Shandong, Jiangsu, Anhui, Zhejiang, Fujian, Jiangxi, Henan, and Guangdong provinces, with a total of 78 spatial units. In 2017, the number of spatial units in the H–H cluster area decreased to 67 but a trend extending from the coast to the inland "axial belt" was observed. The H–H cluster area continued to expand to central and western provinces, such as Hunan, Hubei, Shaanxi, and Shanxi. The conditions of innovation resources, platforms, and the milieu in the above regions were relatively superior, and the strong intercity spatial linkages enabled it to acquire more knowledgeable flows. Hence, the development level of innovation capabilities was higher. Figure 2 shows that the H–L cluster area appeared in the capital cities of the western region, namely, Kunming, Xining, and Urumqi. Most western cities had relatively low levels of innovation capacity, while the province's main innovation factors and social development resources were concentrated in the provincial capital cities, thereby forming a "polarized" differentiation pattern with a high middle and low surrounding. Only Panzhihua city in Sichuan Province had a positive correlation of low innovation capacity. The generally low efficiency of knowledge creation and knowledge flow in the surrounding areas eventually led to Panzhihua becoming a low-concentration depression.

5.1.2. Bivariate Spatial Correlation Distribution Characteristics of Green Technology Manufacturing Efficiency and Innovation Capability

A Geoda 1.14 software (developed by Dr. Luc Anselin and his team, University of Chicago, Chicago, IL, United States) spatial analysis module was used to calculate the bivariate Moran's I value of China's regional innovation capability and green technology manufacturing efficiency in 2011–2017. It explored the overall spatial correlation characteristics and changes between the two elements. The results showed that the bivariate Moran's I estimates were positive during the study period, fluctuating between 0.1034 and 0.1347. With a highly correlative (i.e., passing) significance level test of 0.01, it can be inferred that China's regional innovation capability and green technology manufacturing efficiency had a spatially significant positive correlation.

A bivariate global autocorrelation Moran's I can determine whether an aggregation characteristic of this phenomenon exists in space but it cannot indicate the location of the aggregation exactly. The local spatial correlation characteristics and distribution pattern between the two variables were explored and 2011 and 2017 were selected as the time nodes. The Geoda software was used to conduct a bivariate local spatial autocorrelation analysis of China's regional innovation capability and green technology manufacturing efficiency indicators. The results are shown in Figure 4. The bivariate LISA agglomeration map of China's regional innovation capability and green technology manufacturing efficiency was based mainly on H–H cluster (i.e., core areas) and L–L cluster (i.e., edge districts) spatial units. The spatial evolution characteristics were as follows: the number of core areas rose slowly and the number of edge districts declined rapidly. Specifically, the core areas were distributed in eastern cities, such as Beijing, Yantai, Qingdao, Shanghai, Suzhou, Jiaxing, Shaoxing, and Dongguan, in which urban innovation played a significant role in promoting the improvement of green technology efficiency in the manufacturing industry. Specifically, it was found that innovation factors became increasingly important for improving the green technology manufacturing efficiency in Shandong Province. In contrast, the bottleneck of Guangdong's (i.e., except Shenzhen) manufacturing industry's green technology efficiency was difficult to break through and the spatial core area shrunk gradually due to a series of locking effects that hampered the development path of the manufacturing industry; these effects included an excessive reliance on capital and labor-intensive industries and various problems that hindered the upgrading of the industrial structure, such as the imperfect transformation mechanism of scientific and technological achievements, wasted research and development investment in innovation, lack of core technology, population, and the disappearance of policy "dividends." The fringe areas were distributed in a scattered manner in the northeastern, central, and western regions of China. With time, the fringe areas narrowed gradually in size, shifting from northeast and central areas to the west, indicating that after a period of development in China, the "double-low" areas with a low innovation capability and a low green technology manufacturing efficiency were decreasing, but the effects of innovation in the western region on green technology manufacturing efficiency remained weak.

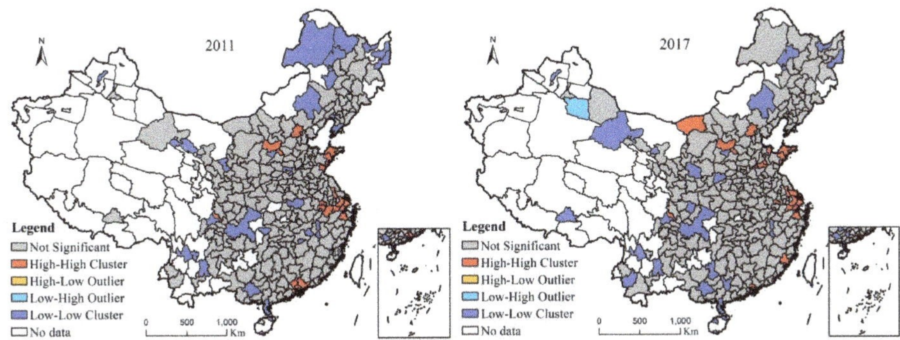

Figure 4. LISA bivariate spatial correlations between green technology efficiency and innovation capability in the manufacturing industry.

5.2. Empirical Analysis and Testing

5.2.1. Benchmark Regression Results

A random effect panel Tobit model regression was used to examine the direct effect of the regional innovation capability on the green technology efficiency of China's manufacturing industry. The results are shown in Table 2.

Table 2. Regression results for the direct effect.

Indicator	(1) Tobit	(2) Tobit	(3) OLS	(4) East	(5) Central	(6) West	(7) Northeast
INN	0.297 ***	0.211 ***	0.182 ***	0.285 ***	0.041	0.188	−0.117
	(0.057)	(0.048)	(0.049)	(0.060)	(0.134)	(0.117)	(0.162)
GDP		0.590 ***	0.239 **	0.651 ***	0.703 ***	0.045	1.074 ***
		(0.113)	(0.110)	(0.192)	(0.198)	(0.262)	(0.255)
FC		0.078 ***	0.040 *	0.052 ***	0.152 ***	0.079 **	0.159 ***
		(0.012)	(0.022)	(0.015)	(0.029)	(0.033)	(0.031)
GI		−0.050 *	−0.049 *	0.037	−0.057	−0.063	−0.078 *
		(0.025)	(0.026)	(0.054)	(0.052)	(0.043)	(0.045)
MS		−1.149 ***	−1.493 ***	−1.260 ***	−1.094 ***	−1.184 ***	−1.081 ***
		(0.099)	(0.094)	(0.157)	(0.226)	(0.187)	(0.248)
Cons	0.159 ***	0.677 ***	1.312 ***	0.606 **	0.642 **	1.170 ***	0.469
	(0.047)	(0.143)	(0.134)	(0.237)	(0.302)	(0.293)	(0.411)
Sigma_u:_cons	0.238 ***	0.280 ***		0.278 ***	0.279 ***	0.301 ***	0.247 ***
	(0.014)	(0.017)		(0.027)	(0.032)	(0.037)	(0.048)
Sigma_e:_cons	0.055 ***	0.045 ***		0.043 ***	0.043 ***	0.048 ***	0.029 ***
	(0.001)	(0.001)		(0.002)	(0.002)	(0.003)	(0.003)
N	887	887	887	371	241	212	63

Standard errors are in parenthesis. Note: *** $p < 0.01$, ** $p < 0.05$, * $p < 0.1$.

In Table 2, only the innovation capability variables were added to column 1 and column 2, hence the result of introducing the control variables of the economic development level, foreign cooperation, government subsidies, and manufacturing scale were exercised. The addition of control variables had little effect on the regression coefficients of the innovation ability variables, which indicated that the research results were robust. The regression coefficients of the innovation capacity variables in columns 1 and 2 were positive and significant, indicating that the overall regional innovation had a significant role in promoting the green technology efficiency of China's manufacturing industry and innovation had become a key power-driver force for improving green technology efficiency in the contemporary era. Therefore, Hypothesis 1 proved to be correct. This conclusion was consistent with the conclusions based on other research fields [49,50]. The regression results of the ordinary least-squares fixed-effect model of the regional innovation capacity and green technology efficiency of manufacturing in Table 2 column 3 also verified the above findings.

From the perspective of regional heterogeneity, columns 4–7 reflect the estimated results of the effects of innovations on the green technology efficiency in the four major sectors of East, Central, West, and Northeast China. The innovation in the central and western regions and the green technology efficiency showed an insignificant positive correlation, while the innovation capacity variable in the northeastern region had an insignificant negative relationship. This result indicated that the innovation capabilities of China's central, western, and northeastern regions had not achieved the effect of promoting green technology efficiency in manufacturing within the study period. The above results corroborated the conclusions obtained in Figure 3 and also proved that Hypothesis 2 was valid. First, the eastern region had strong scientific research strength and a huge technological innovation platform. The scientific research funds and talents were used mostly in the field of experimental development, which was highly relevant to production activities. Therefore, its innovation had a strong spillover bonus to green technology efficiency, which was generally higher than the level of other regions in the country. Second, because of the constraints of the level of economic development of the central and western regions, insufficient investment in scientific and technological innovation, less high-quality knowledge innovation output, and the scarcity of green and efficient high-tech enterprises inhibited the positive effects of innovation on the green technology efficiency of manufacturing. Finally, the northeast region was once China's largest old industrial base but the emergence of the "resource curse" phenomenon caused the northeast economy to show a "cliff-like" downward trend since 2013 [51]. The system and mechanism reform in this area lagged and the manufacturing industry was mostly high-energy-consuming heavy industry, where the development mode was extensive. As a result, one-third of the country's total, i.e., 24 cities, are resource-depleted and face the problem of lagging behind the development of innovative capabilities and green technology efficiency [52].

Among the control variables, government subsidies and manufacturing scale variables deserve further attention. Column 2 shows that the variable sign of the government subsidies was negative and significant at the level of 10%, indicating that the subsidies granted by the Chinese government to manufacturing companies had a significant inhibitory effect on the improvement of green technology efficiency. The reason for this may be that the government subsidies crowded out the private green input of enterprises (crowding-out effect). Enterprises rely heavily on government subsidies, reducing their enthusiasm for green production activities, and the green technology efficiency that was improved by the government investment was offset in whole or in part by the reduction in private investment. In addition, it may also be because China's manufacturing industry is currently at the low-end of the international division of labor. Short-term economic benefits are more attractive than green technology upgrades that have a "large investment and a long payback period." Government subsidies are used mostly to purchase raw materials and expand production, while green technology transformation expenditure accounts for a relatively small amount. China's manufacturing industry has fallen into a path dependence that involves low-tech production models, which hinders the improvement of green technology efficiency. The results in column 7 show that the government subsidies in the northeast region were negatively correlated with the technological efficiency of the manufacturing industry, thereby verifying the above views, while the variables of the government subsidies in columns 4–7 were not significant. From the regression results in columns 2–7, the effects of the manufacturing scale on green technology efficiency were negative, passing the 1% significance level test, and the regression coefficient was significantly higher than for other variables. This finding shows that the larger the scale of manufacturing, the lower the efficiency of green technology in manufacturing. On the one hand, the larger the scale of the enterprise, the lower the production flexibility, and the decreasing effect of scale inhibited the improvement of the green technology efficiency. On the other hand, the extensive development model of high pollution, high energy consumption, and high emissions in China's manufacturing industry restricted the "green" transformation and upgrading of the manufacturing industry.

5.2.2. Mediation Effect Regression Results

This study referred to Wen et al.'s [53] intermediary three-step test method. First, we determined whether the regional innovation capability could promote the improvement of green technology efficiency in China's manufacturing industry. Second, this study examined the role of regional innovation capacity on the intermediary variables. Finally, it explored whether regional innovation capacity and intermediary variables could simultaneously affect the green technology efficiency in manufacturing. According to Muller et al. [54], the existence of intermediary effects must meet the following conditions. First, when no intermediary variables are added, regional innovation capacity has a significant positive effect on the green technology efficiency of manufacturing. Second, this effect is weakened after the intermediary effects are added. Third, regional innovation capacity has a positive effect on intermediary variables. Finally, intermediary variables have a positive effect on the green technology efficiency of manufacturing. Human capital, government revenue, and the waste treatment rate were selected as intermediary variables, and the core explanatory variables and explained variables were consistent with the above. The human capital interpretation index was the number of students in colleges and universities, considering that it will take a certain amount of time for students to convert into labor, i.e., one year from the time of graduation before entering the job market. The statistics of government revenue funds were generally lagging, and thus, we set this indicator to lag one year. The mixed Tobit model was used to test the mediation effect in which the Sobel test results were given (Table 3).

Table 3. Regression results for the mediation effect.

Indicator	(1) Tobit	(2) Human Capital	(3)	(4) Government Revenue	(5)	(6) Waste Treatment Rate	(7)
INN	1.320 ***	0.901 ***	0.684 ***	0.156 ***	1.051 ***	0.102	1.316 ***
	(0.103)	(0.036)	(0.145)	(0.014)	(0.129)	(0.067)	(0.103)
HC			0.676 ***				
			(0.109)				
GR					1.727 ***		
					(0.312)		
WTR							0.034
							(0.042)
Cons	−0.588 ***	0.134 ***	−0.640 ***	−0.031 ***	−0.537 ***	0.847 ***	−0.617 ***
	(0.077)	(0.027)	(0.082)	(0.011)	(0.090)	(0.050)	(0.085)
Sigma:_cons	0.236 ***	0.077 ***	0.225 ***	0.029 ***	0.240 ***	0.154 ***	0.236***
	(0.006)	(0.002)	(0.006)	(0.001)	(0.006)	(0.004)	(0.006)
Obs.	887	717	717	717	717	887	887
Sobel Z		0.0000		4.722		0.6596	
Standard error		0.1002		0.0530		0.0810	
- *P*-value		0.0000		0.0000		0.2316	
Proportion		0.4740		0.1929		0.0025	

Standard errors are in parenthesis. Note: *** $p < 0.01$.

The results in column 1 of Table 3 were similar to the previous findings, and thus, the regional innovation capability had a significant positive effect on the green technology efficiency of China's manufacturing industry. Columns 2 and 3 give the regression results with human capital as an intermediary variable. Column 2 shows that the regression coefficient of innovation capability on human capital was positive and highly significant at 1%, indicating that regional innovation capability significantly promoted the improvement of the human capital level. The regression coefficient of human capital on green technology manufacturing efficiency in column 3 was also significantly positive, which indicated that regional innovation capability can promote the progress of green manufacturing efficiency in China's manufacturing industry through the positive impact of human capital. Specifically, under the condition that other factors remained unchanged, each additional unit of regional innovation capacity directly promoted the green technology efficiency of the manufacturing industry by 0.684 units while promoting the level of human capital by 0.901 units, and thus, the green technology efficiency of the manufacturing industry was increased indirectly by 0.609 units (0.901 ×0.676 = 0.609). The total effect (1.293) was the sum of indirect effect and direct effect, where the indirect effect accounted for 47% of the total effect. Columns 4 and 5 show the estimates of the government's public fiscal revenue as an intermediary variable. The results showed that regional innovation capacity had a positive effect on government revenue. Government fiscal revenue also has a positive effect on the green technology manufacturing efficiency. Compared with column 1, the regression coefficient of the regional innovation capability in column 5 decreased after adding the intermediary variables, thereby indicating that government fiscal revenue was one of the channels for regional innovation capability to promote green technology efficiency in manufacturing. The indirect effect of government fiscal revenue accounted for 19% of the total effect and the intermediary effect was relatively weak compared with human capital. The results of the waste treatment rate as an intermediary variable are listed in columns 6 and 7. Column 6 shows the effect of regional innovation capability on the waste treatment rate and column 7 shows the combined effect of regional innovation capability and the waste treatment rate on the green technology manufacturing efficiency. The results showed that the waste treatment rate did not have a significant effect on regional innovation capability and the green technology manufacturing efficiency, and thus, the waste treatment rate was not a medium for regional innovation capability to promote the green technology manufacturing efficiency. Based on the above results, Hypothesis 3 was partially verified.

5.2.3. Robustness and Endogenous Test

To test the reliability of the above results, i.e., the ability of regional innovation to have a significant positive effect on the progress of the green technology efficiency in China's manufacturing industry, the following robustness tests were conducted. Cities with a special administrative status have comprehensive resource advantages that are unmatched by ordinary prefecture-level cities. Column 1 in Table 4 shows the results of the re-measurement regression after excluding China's municipalities and provincial capital cities. Next, to verify whether the effect of innovation on green production efficiency made a difference with time, column 2 of Table 4 shows the results of randomly deleting the samples of time nodes in 2012, 2014, and 2016 for re-estimation. Then, we investigated whether the core explanatory variables still had a positive effect on the explained variables under different indicator systems. Column 3 of Table 4 shows the effect of replacing the core explanatory variables with urban innovation capability indicators in the "China Urban and Industrial Innovation Capability Report" published by the Industrial Development Research Center of Fudan University in China to estimate the robustness. Finally, outliers and non-randomness affected the regression results adversely. Column 4 in Table 4 shows the results of the core explanatory variables and the explained variables being subjected to the winsorized process of 1% up and down and re-evaluating the regression.

Table 4. Regression results of robustness and endogeneity tests.

Indicator	(1)	(2)	(3)	(4)	(6)	(7)
			Tobit		2SLS	IV–Tobit
INN	0.219 ***	0.529 ***	0.214 ***	0.244 ***	2.327 ***	2.237 ***
	(0.055)	(0.071)	(0.033)	(0.055)	(0.807)	(0.794)
GDP	0.742 ***	1.026 ***	−4.567 ***	0.548 ***	−2.322 ***	−2.231 ***
	(0.122)	(0.145)	(0.144)	(0.114)	(0.846)	(0.815)
FC	0.096 ***	0.029	0.797 ***	0.082 ***	0.072	0.078
	(0.015)	(0.037)	(0.063)	(0.012)	(0.071)	(0.066)
GI	−0.045	−0.094 **	−0.001	−0.047 *	−0.197 **	−0.194 **
	(0.029)	(0.038)	(0.014)	(0.025)	(0.091)	(0.093)
MS	−1.178 ***	−0.966 ***	−0.268 ***	−1.116 ***	3.357 ***	3.391 ***
	(0.121)	(0.143)	(0.067)	(0.098)	(0.187)	(0.191)
Cons	0.507 ***	−0.035	4.131 ***	0.658 ***	−2.035 ***	−2.080 ***
	(0.164)	(0.200)	(0.122)	(0.144)	(0.219)	(0.222)
Sigma_u:_cons	0.258 ***	0.262 ***	0.318 ***	0.278 ***		
	(0.017)	(0.016)	(0.019)	(0.017)		
Sigma_e:_cons	0.046 ***	0.052 ***	0.024 ***	0.044 ***		
	(0.001)	(0.002)	(0.001)	(0.001)		
N	668	505	759	863	887	887

Standard errors are in parenthesis. Note: *** $p < 0.01$, ** $p < 0.05$, * $p < 0.1$. 2SLS: two-stage least-squares, IV: instrumental variables.

The above robust regression results maintained a high degree of consistency with the benchmark regression, where regional innovation capabilities still had a significant role in promoting the progress of China's green technology manufacturing efficiency. Therefore, the conclusions of this study were stable for different model settings and time interval selections. The regression coefficient of the innovation capacity variable in the basic regression was 0.211, and the regression coefficient of the innovation capacity variable after replacing the core explanatory variables with the "China Urban and Industrial Innovation Capacity Report" was 0.214. The comparison found that the regression results were extraordinarily similar, and all passed the 1% significance level test. Consequently, the multi-dimensional and comprehensive index system of the innovation capability level constructed in this study had strong rationality.

Potential sources of endogeneity include missed variable bias. Although we added as many control variables as possible in the model, because the data used in this study were panel data of China's prefecture-level cities and China is a vast territory, certain gaps may exist in the data statistics in which important control variables may be missed. The inclusion of these missing variables in the random error term will cause endogenous problems, resulting in the overestimation or underestimation of the regression coefficient of innovation ability variables. The sources of endogeneity also have mutual causality. It is generally believed that an endogenous relationship exists between innovation and green technology efficiency, i.e., innovation capacity will affect green technology efficiency, whereas regions with a higher green technology efficiency generally have a stronger innovation capacity. Hence, to solve the errors caused by the endogenous problem, the use of the instrumental variable method was tested. The ideal instrumental variable should have a strong correlation with the endogenous variable and must meet the exogenous requirements that are not related to the disturbance term. Geographical indicators are naturally formed and do not affect various indicators in the existing economic system directly, thereby satisfying the exogenous conditions. Lin and Tan [55] used terrain fluctuations as a tool variable for analyzing the effects of the economic aggregation on the efficiency of the green economy. The urban slope is a geographically existing factor that does not affect the green technology efficiency of manufacturing directly but the urban slope is related closely to the urban population, transportation, and other innovative factors. Accordingly, this study used the average slope of the city as a tool variable for innovation ability and used the two-stage least-squares and instrumental variables (IV)–Tobit methods to further identify the causal relationship between innovation ability and green technology manufacturing efficiency.

Column 6 of Table 4 is the result of adopting the two-stage least-squares method with all control variables added. The innovation capability still had a significant positive effect on the green technology efficiency of the manufacturing industry. Moreover, the regression coefficient of innovation ability was improved significantly compared to the estimated value of the benchmark regression coefficient, indicating that benchmark regression underestimated the positive effect of innovation ability on the green technology efficiency of manufacturing. The F-test value of the first stage was 25.84, which was greater than 10, which indicated that there was no weak instrumental variable problem. This study also attempted to use the IV–Tobit method for testing. The results in column 7 of Table 4 show that innovation capability was also conducive to improving the efficiency of green technology in manufacturing.

6. Conclusions and Discussion

Based on the 2011–2017 panel data of prefecture-level cities in China, this study used a spatial autocorrelation method and a Tobit model to analyze the effects of regional innovation capacity on the green technology efficiency of China's manufacturing industry and drew the following conclusions. First, the local correlation pattern of green technology efficiency in China's manufacturing industry was characterized by "large agglomeration and small dispersion." The H–H clusters areas were concentrated in cities along the eastern coast of Beijing–Tianjin–Hebei, Shandong Peninsula, Yangtze River Delta, Fujian Province, and Guangdong Province. The L–H clusters areas were attached around the H–H clusters areas, and H–L clusters areas were scattered in the provincial capital of central cities in the west and northeast. The above conclusion was consistent with that of Wang et al. [56]. Second, Wang et al. [56] viewed the innovation capability gap as an important factor in the imbalance between regions in China. We further found that the regional spatial agglomeration characteristics of China's regional innovation capability were noteworthy, with H–H clusters as the main distribution type. The characteristics of the spatial pattern were distributed as "planar" contiguous patches and the trend inward, extending from the coast to the inland "axial band," appeared gradually. H–L cluster areas all appeared in the capital cities of the western region and only Panzhihua City in Sichuan Province had a positive correlation with low innovation capacity. Third, the regional innovation capacity had a positive spatial correlation with the green technology efficiency of China's manufacturing industry. Specifically, the eastern region had a more prominent role, while the western region had a weaker effect. Fourth, regional innovation had a significant direct promotion effect on the green technology efficiency of China's manufacturing industry. Innovation has become a key driving force in the improvement of green technology efficiency in the contemporary era but the regional differences between the four major sectors were still very significant. The intermediary effect indicated that the regional innovation capacity had a positive effect on the green technology manufacturing efficiency through human capital and government revenue and that the waste treatment rate was not a medium for the regional innovation capacity to promote the green technology manufacturing efficiency. Finally, the various robust regression results maintained a high degree of consistency with the benchmark regression, indicating that the conclusion of this article is credible. After using the instrumental variables, the estimated value of the innovation coefficient of the benchmark regression was found to be improved significantly, and to a certain extent, the positive effect of innovation capacity on the green technology efficiency of the manufacturing industry was underestimated.

The research conclusion of this article has obvious policy implications. First, upgrading the innovation capabilities of the central, western, and northeastern regions is urgently needed. The construction of innovative basic platforms, such as national engineering laboratories and national key laboratories in the central, western, and northeastern regions should be strengthened, along with the dominant position of enterprises in technological innovation. The industrial technology innovation in terms of strategic alliances with universities and research institutes in the eastern region should continue to be built, and the innovation network that collaborates with industry chains across regions should be reinforced to accelerate research and development of key technologies in manufacturing. Policy support for innovation should be increased, intellectual property protection bolstered, further innovation reform programs should be formulated, and the overall innovation ecosystem should be improved. Second, human resource reserves should be consolidated, and priority should be given to the development of education. Basic education should be promoted to improve quality and efficiency, accelerate the development of modern vocational education, and build a new system of industrial talent training. A higher education innovation consortium should be created and support should be provided to local universities and internationally renowned universities to jointly carry out teaching and scientific research activities such that they can strive to cultivate a batch of internationally-renowned strategic scientific and technological talents, as well as high-level innovation-based teams. Finally, the transformation and upgrading of manufacturing green industries should be promoted. The implementation of "Made in China 2025" should be accelerated and energy-saving and environmental protection, new materials, new energy, and other emerging industries should be explored through the use of big data, cloud computing, blockchain, and other new technologies to promote enterprise products to the middle and high ends of the market. Furthermore, high pollution, high-energy-consuming industries should be eliminated gradually by creating an industrial chain of resource recycling in enterprise parks, improving the level of harmless treatment and comprehensive utilization of waste, reducing energy and material consumption, implementing cleaner production supervision, and improving green technical efficiency.

Although some possible problems with the study have been considered, further deficiencies include certain limitations of the data, as well as its focus on only A-share listed companies in the manufacturing industry. As a result, the study does not comprehensively analyze the green technology efficiency of all manufacturing industries in China. Moreover, since 2017, the statistical caliber of the China City Statistical Yearbook has changed, where important factors, such as transportation, population density, and fixed asset investment, were now being considered as control variables. Subsequent research can be carried out from the following aspects: comparing the heterogeneous effects of innovation capabilities on the efficiency of different types of green manufacturing technologies and clarifying the in-depth mechanism of the differences of the effects of regional innovation capabilities on green technology manufacturing efficiency in the different regions in China. This would provide theoretical support for the government to formulate scientific and technological innovation strategies and manufacturing industry planning for the future.

Author Contributions: In this research study, T.W. conceived the study and polished the language; A.S. wrote the introduction and conducted the literature review; Y.F. analyzed the data and wrote the paper. L.W. and

G.T.C. made suggestions and edited the paper. All authors have read and agreed to the published version of the manuscript.

Funding: This research was supported by the National Natural Science Foundation of China (grant no. 41471103) and Major Projects of Philosophy and Social Science Research in Universities of Jiangsu Province (grant no. 2020SJZDA135).

Conflicts of Interest: The authors declare no conflict of interest.

References

1. UNECE. *20 Years of Environmental Performance Reviews: Impacts, Lessons Learned and the Potential to Integrate the Sustainable Development Goals*; Batumi, Georgia, 27 May 2016; Available online: www.unece.org/fileadmin/ DAM/env/documents/2016/ece/ece.batumi.conf.2016.inf.5.e.pdf (accessed on 18 October 2020).
2. Avtar, R.; Tripathi, S.; Aggarwal, A.K. Assessment of energy-population-urbanization. *Land* **2019**, *8*, 124. [CrossRef]
3. El-kassar, A.; Kumar, S. Green innovation and organizational performance: The influence of big data and the moderating role of management commitment and HR practices. *Technol. Forecast. Soc. Chang.* **2018**, 1–16. [CrossRef]
4. Li, Y. Environmental innovation practices and performance: Moderating effect of resource commitment. *J. Clean. Prod.* **2014**, *66*, 450–458. [CrossRef]
5. Schiederig, T.; Tietze, F.; Herstat, C. What is green Innovation? A quantitative literature review. Working Paper 63. In Proceedings of the XXII ISPIM Conference, Hamburg, Germany, 12–15 June 2011.
6. Calza, F.; Parmentola, A.; Tutore, I. Types of green innovations: Ways of implementation in a non-green industry. *Sustainability* **2017**, *9*, 1301. [CrossRef]
7. Fernando, Y.; Wah, W.X. The impact of eco-innovation drivers on environmental performance: Empirical results from the green technology sector in Malaysia. *Sustain. Prod. Consum.* **2017**, *12*, 27–43. [CrossRef]
8. Wang, W.; Yu, B.; Yan, X.; Yao, X.; Liu, Y. Estimation of innovation's green performance: A range-adjusted measure approach to assess the unified efficiency of China's manufacturing industry. *J. Clean. Prod.* **2017**, *149*, 919–924. [CrossRef]
9. Nahm, J.; Steinfeld, E.S. Reinventing mass production: China's specialization in innovative manufacturing. *Engineering* **2012**. [CrossRef]
10. Zhang, D.; Cao, H.; Zou, P. Exuberance in China's renewable energy investment: Rationality, capital structure and implications with firm level evidence. *Energy Policy* **2016**, *95*, 468–478. [CrossRef]
11. Zhang, Y.; Xing, C.; Wang, Y. Does green innovation mitigate financing constraints? Evidence from China's private enterprises. *J. Clean. Prod.* **2020**, *264*, 121698. [CrossRef]
12. Qingqing, W.; He, X.; Yijuan, J. Growing a green economy in China. *Environ. Earth Sci.* **2008**, *12*, 052082. [CrossRef]
13. Kong, T.; Feng, T.; Ye, C. Advanced manufacturing technologies and green innovation: The role of internal environmental collaboration. *Sustainability* **2016**, *8*, 1056. [CrossRef]
14. Zhang, Y.; Sun, J.; Yang, Z.; Wang, Y. Critical success factors of green innovation: Technology, organization and environment readiness. *J. Clean. Prod.* **2020**, *264*, 121701. [CrossRef]
15. Aldieri, L.; Kotsemir, M.N.; Vinci, C.P. Knowledge spillover effects: Empirical evidence from Russian regions. *Qual. Quant.* **2018**, *52*, 2111–2132. [CrossRef]
16. Auteri, M.; Guccio, C.; Pammolli, F.; Pignataro, G.; Vidoli, F. Spatial heterogeneity in non-parametric efficiency: An application to Italian hospitals. *Soc. Sci. Med.* **2019**, *239*, 112544. [CrossRef] [PubMed]
17. Cirone, A.E.; Urpelainen, J. Technovation Political market failure? The effect of government unity on energy technology policy in industrialized democracies. *Technovation* **2013**, *33*, 333–344. [CrossRef]
18. Guo, Y.; Xia, X.; Zhang, S.; Zhang, D. Environmental regulation, government R&D funding and green technology innovation: Evidence from China provincial data. *Sustainability* **2018**, *10*, 940. [CrossRef]
19. Gu, S.; Schwag Serger, S.; Lundvall, B.Å. China's innovation system: Ten years on. *Innov. Manag. Policy Pract.* **2016**, *18*, 441–448. [CrossRef]
20. Gu, S.; Lundvall, B.Å.; Ju, L.; Malerba, F.; Serger, S.S. China's system and vision of innovation: An analysis in relation to the strategic adjustment and the medium- to long-term S&T development plan (2006-20). *Ind. Innov.* **2009**, *16*, 369–388. [CrossRef]

21. Bi, K.; Huang, P.; Wang, X. Technological forecasting & social change innovation performance and in fluencing factors of low-carbon technological innovation under the global value chain: A case of Chinese manufacturing industry. *Technol. Forecast. Soc. Chang.* **2016**. [CrossRef]

22. Gao, Y.; Zang, L.; Roth, A.; Wang, P. Does democracy cause innovation? An empirical test of the popper hypothesis. *Res. Policy* **2017**, *46*, 1272–1283. [CrossRef]

23. The Council on Competitiveness. *Measuring Regional Innovation: A Guide Book for Conducting Regional Innovation Assesment*, 1st ed.; The Council on Competitiveness: Washington, DC, USA, 2005; ISBN 1889866261.

24. Qu, Y.; Yu, Y.; Appolloni, A.; Li, M.; Liu, Y. Measuring green growth efficiency for Chinese manufacturing industries. *Sustainability* **2017**, *9*, 637. [CrossRef]

25. Du, J. Assessing regional differences in green innovation efficiency of industrial enterprises in China. *Int. J. Environ. Res. Public Health* **2019**, *16*, 940. [CrossRef] [PubMed]

26. Hu, S.; Liu, S.; Li, D.; Lin, Y. How does regional innovation capacity affect the green growth performance? Empirical evidence from China. *Sustainability* **2019**, *11*, 5084. [CrossRef]

27. Chen, X.; Liu, Z.; Ma, C. Chinese innovation-driving factors: Regional structure innovation effect, and economic development-empirical research based on panel data development-empirical research based on panel data. *Ann. Reg. Sci.* **2017**, *59*, 43–68. [CrossRef]

28. Li, T.; Liang, L.; Han, D. Research on the efficiency of green technology innovation in China's provincial high-end manufacturing industry based on the RAGA-PP-SFA Model. *Math. Probl. Eng.* **2018**, 9463707. [CrossRef]

29. Motohashi, K. The regional innovation system in China: Regional comparison of technology, venture financing, and human capital focusing on Shenzhen. *RIETI Policy Discuss. Pap. Ser.* **2018**, 18-P-012. Available online: https://www.rieti.go.jp/jp/publications/pdp/18p012.pdf (accessed on 18 October 2020).

30. Yuan, B.; Xiang, Q. Environmental regulation, industrial innovation and green development of Chinese manufacturing: Based on an extended CDM model. *J. Clean. Prod.* **2018**, *176*, 895–908. [CrossRef]

31. Min, S.; Kim, J.; Sawng, Y.W. The effect of innovation network size and public R&D investment on regional innovation efficiency. *Technol. Forecast. Soc. Change* **2020**, *155*, 119998. [CrossRef]

32. Zhang, J.; Chang, Y.; Zhang, L.; Li, D. Do technological innovations promote urban green development?—A spatial econometric analysis of 105 cities in China. *J. Clean. Prod.* **2018**, *182*, 395–403. [CrossRef]

33. Brakman, S.; Marrewijk, C. Introduction: Heterogeneity at different spatial scales. *J. Reg. Sci.* **2009**, *49*, 607–615. [CrossRef]

34. Wei, Z.; Song, X.; Xie, P. How does management innovation matter for performance: Efficiency or legitimacy? *Chinese Manag. Stud.* **2019**, *25*, 275–296. [CrossRef]

35. Miao, C.; Fang, D.; Sun, L.; Luo, Q. Natural resources utilization efficiency under the influence of green technological innovation. *Resour. Conserv. Recycl.* **2017**, *126*, 153–161. [CrossRef]

36. Hou, S.; Yao, M. Spatial-temporal evolution and trend prediction of agricultural eco-efficiency in China: 1978–2016. *Acta Geogr. Sin.* **2018**, *73*, 2168–2183. [CrossRef]

37. Jorgenson, D.; Gollop, F.; Fraumeni, B. *Productivity and U.S. Economic Growth*, 1st ed.; Elsevier: Duivendrecht, The Netherlends, 2016.

38. He, S.; Du, D.; Jiao, M.; Lin, Y. Spatial-temporal characteristics of urban innovation capability and impact factors analysis in China. *Sci. Geogr. Sin.* **2017**, *37*, 1014–1022. [CrossRef]

39. Liu, C.; Gao, X.; Ma, W.; Chen, X. Research on regional differences and influencing factors of green technology innovation efficiency of China's high-tech industry. *J. Comput. Appl. Math.* **2020**, *369*, 112597. [CrossRef]

40. Li, C.; Li, M.; Zhang, L.; Li, T.; Ouyang, H.; Na, S. Has the high-tech industry along the belt and road in China achieved green growth with technological innovation efficiency and environmental sustainability? *Int. J. Environ. Res. Public Health* **2019**, *16*. [CrossRef] [PubMed]

41. Aigner, D.; Lovell, C.A.K.; Schmidt, P. Formulation and estimation of stochastic frontier production function models. *J. Econom.* **1977**, *6*, 21–37. [CrossRef]

42. Battese, G.E.; Coelli, T.J. Frontier production functions, technical efficiency and panel data: With application to paddy farmers in India. *J. Product. Anal.* **1992**, *3*, 153–169. [CrossRef]

43. Yang, Y.; Deng, X.; Li, Z.; Wu, F.; Li, X. Impact of land use change on grain production efficiency in North China Plain during 2000–2015. *Geogr. Res.* **2017**, *36*, 2171–2183. [CrossRef]

44. Zhou, R.; Zhuang, R.; Huang, C. Pattern evolution and formative mechanism of aging in China. *Acta Geogr. Sin.* **2019**, *74*, 2163–2177. [CrossRef]

45. Yuan, H.; Zhang, T.; Feng, Y.; Liu, Y.; Ye, X. Does financial agglomeration promote the green development in China? A spatial spillover perspective. *J. Clean. Prod.* **2019**, *237*, 117808. [CrossRef]

46. Anselin, L.; Syabri, I.; Smirnov, O. Visualizing multivariate spatial correlation with dynamically linked windows. *Urbana* **2002**, *51*, 61801.

47. Xu, D.; Huang, Z.; Huang, R. The spatial effects of haze on tourism flows of Chinese cities: Empirical research based on the spatial panel econometric model. *Acta Geogr. Sin.* **2019**, *74*, 814–830. [CrossRef]

48. Lin, B.; Tan, R. Economic agglomeration and green economy efficiency in China. *Econ. Res. J.* **2019**, *54*, 119–132.

49. Miao, C.; Fang, D.; Sun, L.; Luo, Q.; Yu, Q. Driving effect of technology innovation on energy utilization efficiency in strategic emerging industries. *J. Clean. Prod.* **2018**, *170*, 1177–1184. [CrossRef]

50. Sun, H.; Edziah, B.K.; Sun, C.; Kporsu, A.K. Institutional quality, green innovation and energy efficiency. *Energy Policy* **2019**, *135*. [CrossRef]

51. Fan, J.; Liu, H.; Wang, Y.; Zhao, Y.; Chen, D. "The northeast china phenomenon" and prejudgment on economic revitalization in northeast China: A primary research on stable factors to impact national spatial development and protection pattern. *Sci. Geogr. Sin.* **2016**, *36*, 1445. [CrossRef]

52. Jin, F.; Wang, J.; Yang, Y.; Ma, L.; Qi, Y. The paths and solutions of innovation development in northeast China. *Sci. Geogr. Sinica* **2016**, *36*, 1285–1292. [CrossRef]

53. Wen, Z.; Chang, L.; Hua, K.-T.; Liu, H. Testing and application of the mediating effects. *Acta Psychol. Sin.* **2004**, *36*, 614–620.

54. Muller, D.; Judd, C.M.; Yzerbyt, V.Y. When moderation is mediated and mediation is moderated. *J. Pers. Soc. Psychol.* **2005**, *89*, 852–863. [CrossRef]

55. Lin, B.; Tan, R. Organized crime and business subsidies: Where does the money go? *J. Urban Econ.* **2015**, *86*, 98–110. [CrossRef]

56. Wang, K.; Yu, S.; Zhang, W. China's regional energy and environmental efficiency: A DEA window analysis based dynamic evaluation. *Math. Comput. Model.* **2013**, *58*, 1117–1127. [CrossRef]

Publisher's Note: MDPI stays neutral with regard to jurisdictional claims in published maps and institutional affiliations.

Essay

Energy Re-Shift for an Urbanizing World

Giuseppe T. Cirella [1,*], Alessio Russo [2], Federico Benassi [3], Ernest Czermański [1], Anatoliy G. Goncharuk [4] and Aneta Oniszczuk-Jastrzabek [1]

1   Faculty of Economics, University of Gdansk, 81-824 Sopot, Poland; ernest.czermanski@ug.edu.pl (E.C.); aneta.oniszczuk-jastrzabek@ug.edu.pl (A.O.-J.)
2   School of Arts, University of Gloucestershire, Cheltenham GL50 4AZ, UK; arusso@glos.ac.uk
3   Italian National Institute of Statistics, 00144 Rome, Italy; benassi@istat.it
4   Department of Management, International Humanitarian University, 65009 Odessa, Ukraine; agg@ua.fm
*   Correspondence: gt.cirella@ug.edu.pl; Tel.: +48-58-523-1258

Abstract: This essay considers the rural-to-urban transition and correlates it with urban energy demands. Three distinct themes are inspected and interrelated to develop awareness for an urbanizing world: internal urban design and innovation, technical transition, and geopolitical change. Data were collected on the use of energy in cities and, by extension, nation states over the last 30 years. The urban population boom continues to pressure the energy dimension with heavily weighted impacts on less developed regions. Sustainable urban energy will need to reduce resource inputs and environmental impacts and decouple economic growth from energy consumption. Fossil fuels continue to be the preferred method of energy for cities; however, an increased understanding is emerging that sustainable energy forms can be implemented as alternatives. Key to this transition will be the will to invest in renewables (i.e., solar, wind, hydro, tidal, geothermal, and biomass), efficient infrastructure, and smart eco-city designs. This essay elucidates how the technical transition of energy-friendly technologies focuses on understanding the changes in the energy mix from non-renewable to renewable. Smart electricity storage grids with artificial intelligence can operate internationally and alleviate some geopolitical barriers. Energy politics is shown to be a problematic hurdle with case research examples specific to Central and Eastern Europe. The energy re-shift stressed is a philosophical re-thinking of modern cities as well as a new approach to the human-energy relationship.

Keywords: rural-to-urban transition; energy mitigation; urbanization agenda; smart city; energy landscape; urban energy transition; alternative energy technologies; sustainable energy; geopolitical energy change; Central and Eastern Europe

check for updates

Citation: Cirella, G.T.; Russo, A.; Benassi, F.; Czermański, E.; Goncharuk, A.G.; Oniszczuk-Jastrzabek, A. Energy Re-Shift for an Urbanizing World. *Energies* **2021**, *14*, 5516. https://doi.org/10.3390/en14175516

Academic Editor: Donato Morea

Received: 23 July 2021
Accepted: 31 August 2021
Published: 3 September 2021

Publisher's Note: MDPI stays neutral with regard to jurisdictional claims in published maps and institutional affiliations.

1. Introduction

The year 2007 marked a fundamental phase in human history: half of the world, for the first time, became urbanized [1]. A change so important that it has been juxtaposed with other fundamental steps in our evolution, such as the agricultural and industrial revolutions, thus coining a new anthropological term, homo urbanus [2]. It is a development that does not seem destined to stop, since the latest available data indicate that in 2018 the percentage of urban population (i.e., globally) has risen to 55%, a share that will exceed, according to official forecasts, 68% in 2050 [3–5]. An even more radical change if we consider that in 1950 the percentage of rural population was 70%. It is not by chance, therefore, that cities have progressively assumed a more central role in the economy, aided by the process of economic globalization [6], and that they represent, for some, the greatest human invention [7–9]. Urbanization, in fact, is not new but rather a process that has accompanied human beings since their birth, significantly characterizing their social and economic circumstances [10]. However, it is only since the nineteenth and twentieth centuries that we can speak of "urbanized societies" [11]. The shift from rural to urban has been accompanied, in today's advanced development economies, by the shift from societies characterized by high levels of birth and mortality to societies in which these two endogenous forces are both at low levels. These profound transformations, which in many developing countries are still in progress, are grounded in the theory of demographic transition [12] and that of mobility [13]. Together with these transitions, economic and social development has taken place, i.e., the so-called modernization era [14]. The interconnections between rural-to-urban transition and economic development are well known; e.g., in 2015, 80% of the global gross domestic product (GDP) was generated in cities [15]. These are even clearer if we think of the deep differentiations that exist, and will persist, between advanced and developing countries, precisely in relation to urban population growth processes [16]. According to official data from the United Nations, in 2018 the percentage of urban population was about 79% in more developed regions (MDRs), while it was almost 51% in less developed regions (LDRs); in 1950, these shares were 55% and 18%, and it is estimated that they will become about 87% and 66% in 2050, respectively [1]. The world is and will thus inexorably be increasingly urban. There is therefore a need to govern this transition constructively and sustainably by not allowing such processes to rule over us.

Historically, the driving force behind urbanization has been migration. In fact, urban population growth is achieved mainly through migration mechanisms, i.e., the different capacity of cities to attract non-urban populations, rather than on the different levels of natural growth of urban and non-urban populations [17]. From a theoretical point of view, the process of urbanization and, more generally, of urban development can be seen as a combination of incoming and outgoing movements. The former is defined as centripetal and concerns rural-urban migration, gentrification, re-urbanization, and urban sprawl renewal. In contrast, the latter is known as centrifugal and refers to suburbanization, urban sprawl, and counterurbanization. Of course, these types of categorizations can vary greatly if we consider the level of development of observed countries and regions [18]. But what will this inexorable process of urbanization mean for the environment, for health, for sustainability, and in terms of energy? Opinions are varied and much will depend on which energy and environmental profiles cities associate themselves with. It is a question of considering urban growth as an opportunity and not as a cost. At present, especially in LDRs, migration towards cities along the rural-urban axis is interconnected with environmental processes, such as desertification and climate change, that these same megacities and large conurbations help to reinforce through the pollution they cyclically produce and trigger. Moreover, if the increase in cities could determine positive externalities [15], the negative externalities that deregulated urbanization processes—defined in the literature as spontaneous, at best, and wild, at worst—can determine the health, urban microclimate, and general conditions of human habitation of urban environments [19,20].

A relevant issue is obviously the energy dimension and its design and management with respect to urban development, both in progress and, above all, in the future, which, as we have pointed out, will largely impact LDRs. Many contributions have been devoted to this issue [21–23]. In particular, in a recent paper that appeared in the UN Chronicle by Philipps and Smith [24], it is clearly stated that "sustainable urban energy is the future." They argue that the implementation of renewable energy strategies in urban environments has become energy impartial, e.g., through a change of sources, but also by ensuring they are actually sustainable and cost-effective. In particular, the point that needs to be emphasized is that the next, inevitable, phase of urban growth, especially in LDRs, must favor processes of environmental sustainability, thus interrupting the vicious cyclic processes that are not sustainable in the medium to long term. The key points, in this regard, on which to invest resources should include solar power, efficient infrastructure, and eco-cities. The process of urbanization is challenging, but we believe it is also an opportunity. From this point of view, cities, and as such by urbanization, can represent advantages [25] for urban environmental services via effective energy mitigation strategies. This essay expands on the notion of an urbanizing world and focuses on three points of understanding this phenomenon: (1) internal (i.e., from a smart city design and techno-innovative perspective),

(2) technical (i.e., from a mitigative transitory outlook), and (3) geopolitical (i.e., from our ability to change and get along as nation states using Central and Eastern Europe as case research).

2. Materials and Methods

This is an expository essay that provides a focused explanation of the energy re-shift for an urbanizing world and correlates this viewpoint with urban energy needs. The methodology for the essay is desk-based research. Data were collected on the use of energy in cities and, by extension, nation states within the context of the essay's three subject matters—i.e., internal urban design and innovation, technical transition, and geopolitical change—over the last 30 years (1991–2021). Urban-centric data focused on smart city design techniques, energy alternatives, and renewables. Specific case research from Central and Eastern Europe is used to stress the economic re-shift of energy markets and their potential for energy poverty and geopolitical shifts. We analyze how these differences affected energy development and sustainability in the context of the rural-to-urban transition. After which, the main challenge is to consider, in the short and medium terms, if the current policies for energy development and use are viable in an urban and regional context.

A scoping literature search was completed using the following electronic resources: Web of Knowledge, Scopus, Science Direct, Google Scholar, and Google. The study synthesized exploratory keywords aimed at mapping key concepts, types of evidence, and gaps in the research by systematically searching, selecting, and synthesizing existing knowledge. Explanatory keywords were derived by using a combined star busting [26] and brainstorming [27] approach, as well as the stepladder method of accumulating additional keywords as research was found [28] (see Appendix A for a list of keywords utilized in the analysis). The literature was compiled and publications were systematically analyzed using strategic and critical reading methods [29,30], as presented in Table A1. We identified more than 3000 articles and grey literature in the first step of the search. To better focus on the essay's three subject matters, we filtered out literature published before 1991 and omitted literature discussing rural-related research as well as narrow technological and engineering-based perspectives, leaving approximately 200 publications in the form of books, scientific articles, and technical reports. Corresponding references, cited in the text, are detailed as data sources and based on the analyses and know-how of the scientific experts of each studied theme. Based on all these materials, a set of proposals to develop internal, technical, and geopolitical backing was formulated to uphold the urbanization agenda and to mitigate energy sustainability strategies.

3. Results and Discussion

3.1. Smart City Landscape Design and Energy Innovation: Internal Urban Strategies

To solve the two worldwide environmental concerns of urbanization and rising carbon dioxide (CO_2) emissions, a smart and urban energy transition is needed [31]. The "smart" approach to urbanism and city development has sparked controversy in sectors such as engineering, innovation, and the social sciences [32]. While sustainability is not often a primary goal of local smart city implementation, the smart agenda raises the bar for achieving energy sustainability goals [32]. Hence, the energy system is one of the most important components of a smart city, as it plays a critical part in the transition to a more sustainable urban lifestyle [33]. The use of renewable energy sources is shown to make a major contribution to lowering pollutant emissions and improving living environment quality [33]. A variety of components and environmentally friendly elements must be integrated for cities to have smart energy systems. The energy systems that provide these cities with heating, cooling, and electricity must be clean, renewable, and abundant, as well as efficient, effective, and secure [34]. However, smart cities are rarely discussed in academic research in the fields of landscape architecture, urban design, and planning [35]. The main function of urban and landscape design in the building of the smart city is based on the integration of technology components with the physical city, including residence and public spaces, politics, economy, and ecology, among other things [35]. The spatial patterns of urban energy systems evident in the built environment are represented by urban energy landscapes [31]. In urban energy landscapes, spatial regularities in the organization of energy provision and consumption systems are visible [31]. The built environment's architecture, as well as people's views of technology, may influence how much energy is used [31]. In this context, renewable energy and green infrastructure will play a crucial role in the development of smart and sustainable cities [36]. However, wind turbines and solar panels are not enough to make the shift from fossil fuel to renewable energy [37]. Landscape design, for instance, can help to reduce periodic variations in energy supply as well as low energy density, which are two of renewable energy's main drawbacks [37]. Energy-conscious planning and design may also increase the efficiency of available energy, whether it comes from renewable or non-renewable sources [37]. As such, a well-planned landscape saves energy and can pay back the initial expenditure in as little as eight years [38]. For example, by allowing winter sun in, an eight-foot deciduous tree may save hundreds of dollars in air conditioning expenditures while also lowering heating and lighting costs [38]. Energy conservation not only saves money but also benefits the environment by reducing the usage of natural resources [38]. Specific landscape design methods are determined by regional climatic conditions as well as the microclimate immediately around buildings [38].

Studies have shown that green infrastructure, which may be built or restored, might help to reduce an area's overall energy demand and, in doing so, help to mitigate the "urban heat island" effect [39]. By shading building surfaces, deflecting solar radiation, and releasing moisture into the environment, trees, green roofs, and other green infrastructure features can help to cool metropolitan environments [39]. In contrast to buildings without trees, a building with trees can consume 2.3% to 90% less cooling energy and 1% to 20% less heating energy due to windbreak effects [40]. In particular, shade from trees has a greater cooling impact than evapotranspiration from lawns, resulting in significant cooling load reduction [41]. Over 15 years, McPherson and Simpson [42] predicted that planting 50 million trees to shade the east and west sides of residential buildings in California would reduce cooling by 1.1% and peak load demand by 4.5%. Moreover, green walls, green facades, and green roofs are examples of exterior greenery systems which can save energy as well as benefit the built environment [43,44]. In Arizona, Yuan and Rim [43] found that green walls can save up to 27,000 kWh/y at a primary school in Phoenix while a green roof can save up to 69,000 kWh/y. In China, Tan et al. [45] investigated the energy-saving potential of building envelope integrated green plants (BIGP) in hot summer and cold winter climates, using comparison tests between a vertical greening room and a reference room. During the winter, BIGP reduced the heat flux density of the outside wall by 3.11 W/m^2, while the reference room's hourly power usage remained 1.22 times higher. BIGP conserved energy at a rate of about 18%. During the summer, the heat flux density of the reference room's external wall was 4.15 W/m^2 higher than that of the vertical green room, with the hourly power usage being 1.33 times higher. It can be stated that BIGP saves roughly 25% of the energy it consumes. Modeling findings evaluated the cooling advantages of green areas in proportion to the mean height of buildings on Gulou Campus in Nanjing, China, yielding 5.2 W/m^2 of cooling energy and saving 1.3×10^4 kW/h over a single daytime hot summer period, according to research by Kong et al. [46].

Green infrastructure can also be promoted as a low-cost strategy for mitigating the carbon footprint of industrial energy. For example, green infrastructure in Hangzhou, China offsets 18.57% of the carbon produced by industrial businesses each year through sequestration and stores an amount of carbon equivalent to 1.75 times the yearly carbon emitted by industrial energy consumption within the city [47]. Green infrastructure may also be a solution to several water-related issues originating from the energy sector. It may be used to enhance the energy efficiency of power generation, in addition to conserving the water environment and water supplies in general [48]. Green infrastructure, when combined with grey infrastructure, such as hydroelectric dams, can increase its lifetime and efficiency [48]. In addition, green infrastructure can be used for the production of biomass energy, even though it has a greater spatial footprint than other energy carriers (e.g., solar power). Important research is lacking on whether a significant increase in biomass use in cities is feasible [36]. Biomass can be used to produce energy from green infrastructure pruning and urban agriculture (e.g., via edible green infrastructure) but requires a substantial amount of (prolonged) maintenance. For example, in Milan, Ferla et al. [49] assessed the biomass of urban greenery maintenance from an energy perspective. They found a biomass potential energy between 26 and 76 GWh, relative to the green-based inventory, regulation, plan, and informatic system used to assay the research [49].

A good example of integration of green infrastructure and renewable energies is the Beddington Zero (fossil) Energy Development (BedZED) in the London Borough of Sutton, United Kingdom (Figure 1). Its major accomplishments include the integration of infrastructure systems for synergistic efficiencies, the installation of renewable energy infrastructure (e.g., from the sun and wind), and the closure of various energy and water loops [50]. In their analyses, the planners and designers of these infrastructures took into account the life cycles of various systems and processes [50]. BedZED was designed to have a low environmental impact during construction and operation, allowing residents to live within their fair share of the earth's resources. Hodge and Haltrecht [51] point out some of the key operational goals of the design:

- reduce water consumption by 33% compared to the national average,
- reduce electricity consumption by 33% compared to the national average,
- reduce space heating requirements by 90% compared to the national average,
- reduce private fossil-fuel car mileage to 50% of the national average, and
- eliminate CO_2 emissions from energy consumption [51].

Figure 1. BedZED project in Hackbridge, London borough of Sutton, with solar panels, wind cowls, and green roofs. Source: Image credits Tom Chance https://www.flickr.com/photos/tomchance/10 08213420/ CC BY 2.0, accessed on 22 July 2021.

BedZED households use 2579 kWh of electricity per year, which is 45% less than the Sutton average. BedZED uses gas to power the district's heating system when the biomass combined heat and power (CHP) plant is not in use. Households use 3526 kWh of heat (e.g., from gas) per year on average, which is 81% less than the Sutton average [51]. The prototype CHP unit from BedZED was designed to be completely automated, with unattended start-up and shut-down and stringent, automatically controlled operational parameters. The plant is planned to work 24 h a day, seven days a week; however, because of noise limits of 37 dBA at 20 m, it only runs for 18 h per day [51]. In addition, on the rooftops and in the south-facing second floor windows, there are 777 m^2 of photovoltaic panels made up of 1138 laminates as well as garden roofs that provide several ecosystem services [51].

The European Green Deal (i.e., the European Union's most significant step toward climate neutrality, issued at the end of 2019) as well the recent European COVID-19 Recovery Fund, represent an opportunity to accelerate the development of renewable energies and green infrastructure-based projects in European cities [52]. Moreover, smart information and communication technology, in combination with urban green infrastructure planning concepts, could be a powerful instrument for coordinating and managing energy issues and the bettering of sustainable smart cities [35]. To navigate this, there will be the need for a technical transition within the urban energy grid that focalizes on non-renewable to renewable forms, something that is fast occurring in much of the MDRs of the world.

3.2. Technical Transition of Energy-Friendly Technologies: Urban Energy Mitigation from Non-Renewable to Renewable

The path of sustainable development will certainly lead to changes in the energy mix from non-renewables towards 100% renewables (i.e., solar, wind, hydro, tidal, geothermal, and biomass). On land and specifically in urban centers, this is possible with the development of appropriate models that are based not on generating facilities and a distribution network, but on storage systems and feedback loops into the distribution grid at times of temporary reduction in supply. A far-reaching concept is the use of smart grids (i.e., defined as artificially intelligent) and advanced power grids that are sustainability-oriented and energy efficient within a smart city design [53]. This type of urban energy landscape will internationally distribute and store electricity, and balance production and transmission volumes in relation to consumption needs [54]. To date, this concept exists in terms of electricity storage based on different types of battery and capacitor systems, even though at present the technology bears limited capacity and limited lifetime. Energy consumption itself varies greatly in space and time (e.g., on a daily, weekly, and annual basis), and does not coincide with any potential seasonality of renewable energy production. As a result, differences should be covered by energy storage systems [55–57] until better technological innovation can be achieved. Currently, known technologies require the use of lithium and cobalt and, as of 2021, the resources of these elements (i.e., in the form of various types of minerals) are not sufficient to cover the demand for the production of batteries for all-energy storage needs [58].

Another challenge is low resilience (i.e., intermittence and unpredictability) of solar and wind energies, which are not available every day or hour. These alternatives are also vulnerable to changes in weather conditions and, in the case of solar, after the complete life cycle of solar modules environmental, safety, and health concerns arise via their disposal [59]. Even though these and other challenges constrain the development of alternative energy-maintaining fossil fuels as the lead energy source worldwide, technological breakthroughs that sharply reduce the cost of specific investments can significantly accelerate the pace of their development and advancement into the main power grid. However, it should be stated, there is no lack of energy from renewables nor are they in short supply. Solar power alone could produce 3.1×10^{17} kWh per year with an annual global energy demand of 1.6×10^{14} kWh [60–62]. Moreover, the solar redesign of cities could include the expansion of solar panels to other parts of the cityscape other than the tops of buildings, e.g., solar pavement [63,64], solar windows [65,66], and solar farms located in and out of city limits (as found throughout many parts of Spain [67]). Currently, the world's largest renewable energy hub, the Western Green Energy Hub (WGEH) in Western Australia, plans to mix solar and wind and is poised to match Australia's entire energy fleet, contributing a whopping 50 GW in power generation. WGEH is planned to be finalized by 2028 and is set to cost USD 95 billion, spanning 15,000 km^2 [68].

In addition to the issue of electricity storage, alternative fuels such as liquid hydrogen or methanol can be used as energy carriers. Both of these fuels can be produced using conventional electricity. However, their use in the context of sustainable development implies that they should be produced using only renewable sources. To put it simply, producing hydrogen requires water and electricity for its electrolysis, whilst producing methanol requires CO_2 (which human beings are increasingly supplying) and electricity. Regardless of the process of transforming the sources of energy consumed in cities, there is a clear process for reducing energy demand. Since 75% of the global greenhouse gas (GHG) emissions are sourced from the urban landscape, urban energy mitigation has become a serious part of the emission reduction process. Urban energy mitigation is complex. It is entrenched in two basic principles: (1) reduce the demand for energy by consumers and (2) change the energy to a clean-oriented source. This is noted, for example, in the recent changes to the German Federal Ministry for Economic Cooperation and Development's [69] report "Climate Change Mitigation in Cities: Urban Action to Reduce Greenhouse Gas Emissions," in which six sectors of urban energy mitigation exist: urban planning, buildings, transport and mobility, energy, waste, and water and sewage management. From this report, buildings (i.e., commercial, institutional, industrial, and residential) make up a 63% share of GHG emissions, with transport and mobility second at 28% (Figure 2). Another comprehensive study delivered by the consortium of the World Resources Institute, C40 Cities Climate Leadership Group, and ICLEI–Local Governments for Sustainability [70] defines these six sectors as: stationary energy; transportation; waste; industrial processes and product use; agriculture, forestry, and other land use; and any other emissions occurring outside the geographic boundary as a result of city activities. Regardless of the classification, it is crucial to develop and disseminate measures, actions, and solutions that lead to the highest possible reduction in urban emission pollution. This practice, however, is spatially limited and dependent on a number of local factors and conditions. The most frequently mentioned include:

- land use limitation for urban purposes by smart and compact designing of the public space,
- resource efficient modernization and new construction of buildings towards green building and climate-neutrality,
- smart and sustainable urban mobility leading to a more sustainable share for pedestrians and shared mobility users as well as public transport passengers, in return for a reduction in car users [28,71],
- postulated decentralization of renewable energy supply (i.e., mainly based on solar photovoltaics),
- improvement of solid waste and water and sewage management (i.e., the 3R strategy: reduce, reuse, and recycle) or a complementary waste-to-energy model-based city economy, and
- reviewing the energy mix while increasing the share from renewable sources.

To better understand the technological energy transition argument (and some would say urgency), emitted pollutants from cities are a driving force. Currently, the largest amount of potential reduction in GHG emissions comes from the energy sector in cities, i.e., 46% of all GHG emissions, wherein two-thirds is dedicated to electricity generation and the other third to fossil fuel extraction [72]. For example, within the transport sector, GHG emissions are estimated at 70% from road, 20% from aviation, and 10% from shipping [72]. Since these figures provide valuable insight into the city-to-GHG emissions breakdown, solutions should start with or, at the minimum, include the reduction in emitted pollutants within the mitigatory process. In order to achieve GHG emissions reduction via energy mitigation, many cities worldwide have already introduced a wide range of solutions and measures related to all aspects of a city life. Some noteworthy urban energy mitigation-based studies include: (1) the United Nations Framework Convention on Climate Change's (UNFCCC) [73] report on the urban environment with related mitigation benefits and co-benefits of policies, practices, and actions; (2) the Intergovernmental Panel on Climate Change's [74] report on human settlements, infrastructure, and spatial planning; (3) C40 Cities et al.'s [72] report on the future of urban consumption in a 1.5 °C world; (4) Santamouris et al.'s [75] research on heat mitigation technologies to improve sustainability in cities; (5) Santamouris et al.'s [76] work on the energy impact of urban heat island research in terms of climate and energy potential of mitigation technologies; (7) Zawadzka et al.'s [77] assessment of the heat mitigation capacity of urban greenspaces; and (8) Carbfix's [78] research outside of Reykjavik, Iceland, that can turn CO_2 into stone [79]. The issue of urban energy mitigation shows a crucial relationship between the GHG emission reduction goal, which is key at the national and international level for determining which of the developed and implemented measures are subject to international legislation, and cities who, as major GHG emission contributors, are responsible for municipal scale activities introduced by city authorities. In combination, close cooperation between individual cities (often via umbrella organizations such as the C40 Cities Climate Leadership Group and ICLEI–Local Governments for Sustainability) and global bodies such as the United Nations can parallel their efforts, or work together, on the sustainable energy agenda (e.g., the City of Vancouver, Canada, Climate Emergency Action Plan [80] that resembles the United Nations Sustainable Development Goal 11 to make cities and human settlements inclusive, safe, resilient, and sustainable by 2030 [81]). Moreover, the UNFCCC Non-State Action Zone for Climate Action gathered information on 2578 cities from 118 countries representing 10.2% of the global population, and found that using energy initiatives as presented in Table A2 would reduce approximately 2.8 Gt of CO_2 emissions by 2050 [82] (Appendix B).

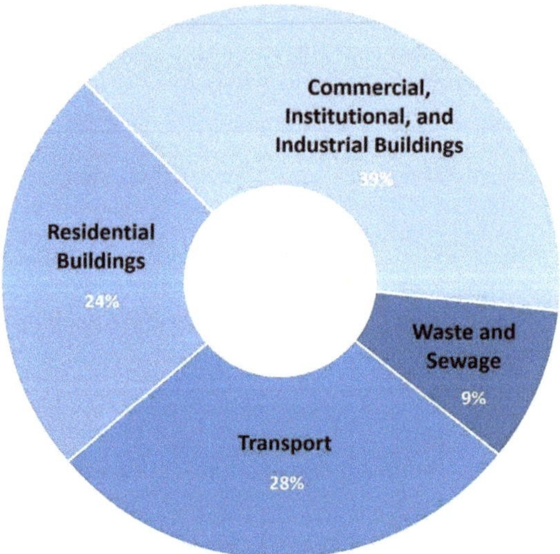

Figure 2. Share of GHG emissions in cities by sector, 2018. Source: German Federal Ministry for Economic Cooperation and Development [69].

The intricate phases of transitioning from non-renewable to renewable sources of energy are complex and no fixed solution is available. As noted, there are important city and international strategies that can aid in this ongoing transition. One important strategy worth noting is the cross-city partnership scheme, e.g., the Covenant of Mayors initiative, which interlinks thousands of municipalities who voluntarily commit to implementing European Union climate and energy objectives [83]. This urban energy mitigation initiative noticeably starts at the city scale, however, there is a global dialogue, i.e., with interlinking facets between international organizations, central governments and municipalities, and industry representatives, that can directly affect any country's energy mix. The elephant in the room we are referring to is the geopolitics of energy and its availability and distribution worldwide. Energy politics has increasingly come to a crossroads with the way cities and countries interact with each other and how citizens view potential energy needs.

3.3. Geopolitical Energy Management and the Economic Re-Shift: Central and Eastern Europe Case Research

Countries have been jostling and fighting over borders for as long as cities have existed. Civilizations have come and gone and formulate the precedent for how cities have developed. Today's geopolitical chessboard is very much rooted in this historical make up. Over the last decade, geo-economics and geopolitical processes are increasingly influencing energy management in various parts of the world. Military conflicts in the Middle East (e.g., Syria and Iraq [84] and the Arab-Israel conflict [85]), the crisis in Venezuela [86], China's growing energy demands [87,88], and even the almost harmless blockade of the Suez Canal in March 2021 [89] have led to increased risks and higher energy prices. These price fluctuations are all interconnected to the energy politics of the day. For example, the linkages and energy collaboration between Iran [90], Russia [91], and Central Asian countries [92] can currently only partly satisfy the enormous energy appetite of China's growing economy [93], even with the outbreak from the COVID-19 pandemic. For example, in Iran, its primary barriers to solar energy development still remain economic, since it still faces heavy international sanctions in collaboration with the low price of fossil fuels [94]. In this case, the levelized cost of solar energy is still much higher than conventional technologies for electricity generation [95]. Moreover, the United States, with the transition from the Trump to Biden Administration, has somewhat eased pressure on Russia in terms of energy influence in the European Union [96]. This is expressed in a more loyal attitude from American authorities towards the completion of the construction of the Nord Stream 2 pipeline, a second Baltic Sea pipeline connecting Narva Bay, Russia, and Greifswald, Germany (Figure 3) [97]. Under the influence of such global shifts, the authorities of some of the most vulnerable countries, e.g., Ukraine, Poland, Slovakia, Belarus, and the Baltic states, are forced to revise their energy policies. A brief examination of the first two countries listed will showcase important examples of economic re-shifts in the region and exemplify the complexities of energy poverty at the city level and hard power at the international. The geopolitics of energy is obviously a vast and complex issue. Clearly, energy politics touches every country with no single solution. The examples used from Central and Eastern Europe exemplify the complexity and individuality of the energy problem each country faces, i.e., in securing its energy needs and ability to maneuver on the global chessboard between economics and energy resources.

Figure 3. Nord Stream 2 pipeline. Source: Nord Stream 2 [98].

3.3.1. Ukraine: Energy Hardship

The launch of the Nord Stream 2 pipeline will leave the composition of natural gas suppliers to the European Union market essentially unchanged [99]; however, it will significantly weaken the position of Ukraine and its allied neighbors in energy and political independence from Russia. Furthermore, the completion of Nord Stream 2 may challenge the European Union's principles in terms of solidarity and trust [100], complicating the energy policy and management in Central and Eastern Europe and policy relations inside the Union itself. Central and Eastern European countries have an acute problem of energy poverty, in which about a fourth of the population is exposed to hidden energy shortages [101]. This is especially true in the case of Ukraine. This problem is sometimes exacerbated by the inadequate actions of the authorities at all levels of government. In Ukraine, where the problem of energy poverty has worsened due to the COVID-19 pandemic, the cost of energy resources in 2020 (despite being slightly economically cheaper) was that they were harder to utilize [102,103]. The expansion of the European Union's energy poverty policies, i.e., fueled by clean energy for all Europeans, as well as regulations and the creation of the European Union Energy Poverty Observatory and the ENGAGER project [104,105], has begun to gradually address this problem, but there is still much to be done. An important factor for sustainability of energy management is alternative energy that can be diversified by varying countries, especially at the city level. Throughout much of Central and Eastern Europe, but especially in Ukraine, cities have been struggling to resolve a number of these barriers (i.e., economic, legal, sociocultural, technical, etc.) that can prevent prompt energy growth and development. In Ukraine, the share of renewables within the total energy mix is still very small. The continual dependence on traditional forms of energy interlocks its growing urban population with noticeable energy blackouts and shortages that hamper its development. Ukraine, as well as other Central and Eastern European countries, looks to the case of Poland as a prospective country that leads the region, as its energy industry has positively undergone comprehensive change in respect to the European Union's energy regulations, influencing energy transition towards climate neutrality, the aging coal stock of Poland's generating units, and increasing environmental awareness society-wide.

3.3.2. Poland: Energy Advancement

In Poland, the need to increase energy efficiency, in recent years, has resulted in energy needs going up—this has been simultaneous with climate change and the growth of customer preferences for energy-efficient products [106]. These changes clearly reflect the European Union and national regulatory bodies and legal stances currently pressuring authorities. An important document relevant to energy efficiency is the European Union Directive 2006/32/EC, which specifies energy end-use efficiency and energy services [107]. Its provisions were ratified on 1 January 2008, and it was deduced that Member States would achieve a reduction in energy consumption of 9% compared to their baseline level between 2008 and 2016. The requirements of Directive 2006/32/EC imposed an obligation to develop and prepare national plans to achieve these objectives, i.e., Energy Efficiency Action Plans (EEAPs) based on the assumptions of the European Union energy policy [108]. Poland's EEAP includes a description of energy efficiency measures by end-use sectors and calculates final energy savings as an ongoing metric. This plan was developed based on the Energy Efficiency Act of 15 April 2011 [109], which sets out the legal framework for improving energy efficiency, i.e., taking into account the leading role of the public sector and establishing supporting mechanisms and systems for monitoring and collecting the necessary data.

Another important document is ISO 50001, i.e., the guidelines of the energy management system requirements, which was adopted in Poland in 2012 and regulates energy use and consumption according to defined energy usage indicators [110]. In parallel, yet another document adopted the same year was the Directive 2012/27/EU of the European Parliament and of the Council of 25 October 2012 on energy efficiency that established a common framework to reduce primary energy consumption in the Union by 20% [111]. This was an important factor for the success of developing and reaching the European Union's energy strategy by 2020 [111], compelling Poland to achieve an energy savings of 17% in that year. As a result, its implementation led to the National Law of Energy Efficiency Act of 20 May 2016, imposing widespread energy reform and a national rethinking. Another European Parliament Directive 2018/2002 of 11 December 2018 amended the Directive 2012/27/EU on energy efficiency by introducing an energy efficiency improvement target of 32.5% by 2030 [112]. This directive further focalized Poland's energy sector into passing the Compensation System Act of 2019, which introduced a new support system for entities performing business activities in energy-intensive sectors and relating subsectors. These reforms included compensation for transferring the costs of purchasing emission allowances (i.e., within the meaning of the Law on the Greenhouse Gas Emission Trading Scheme) for the price of electricity consumed to produce products by such entities [113,114]. At length, this energy advancement highlights a roadmap for other Central and Eastern European countries and firmly presents Poland's ability to diversify its energy use, preventing it from becoming dependent on any one country or energy source (Figure 4) [115].

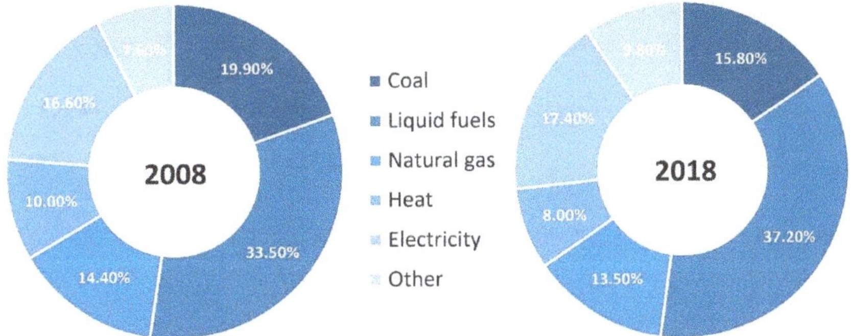

Figure 4. Structure of final energy consumption in Poland by carrier, 2008 and 2018. Source: Peryt et al. [115].

The last decade has brought significant changes and progress in improving energy use in Poland. The energy intensity of its GDP has decreased by nearly 30%. The greatest impact this has had is on its ability to optimize urban and industrial processes, modernize its lighting systems, and initiate thermomodernization projects throughout the country [108]. The application of effective energy management, specific to city development, still has a number of barriers, some being the lack of widespread use of integrated design of buildings (i.e., from investment) and the implementation of fragmented legal regulations. Other barriers concern the lack of an effective system of financial support for investors to viably participate in the energy-efficiency market, since enterprises and residential construction can lack know-how and awareness of modern technologies. Energy management systems, in accordance with the principles of sustainable development, have had to overcome a number of improvements relating to quality of work, life, and the environment, while still maintaining viable economic outputs [116]. Among the instruments used to improve energy efficiency, horizontal measures include sector-based instruments that regulate energy standards within manufacturing, construction, transport, government, and households. Market mechanisms, which include a system of white certificates, have been implemented to increase energy efficiency in energy generation, transmission, and consumption processes [117]. As such, the market motivator for taking energy efficient action has had a trickle-down effect on consumer behavior, in which end users are more aware of energy consumption and the factors that influence quantity and quality of energy usage [118].

In terms of renewables, Poland, generally, has stood out in high support for the development of these types of energy sources. As many as 87.3% of Poles support financial support of the state for the creation of new renewable energy production, with only 6.6%

in opposition [119]. This support highlights the newly implemented Energy Policy of Poland until 2040 (PEP2040), of which renewable energy sources make up an important component in achieving climate neutrality. On 2 February 2021, the Council of Ministers approved PEP2040 as a new strategy setting a new standard for the development of the sector. PEP2040 directs strategic investment decisions aimed at geopolitically leveraging the economy, raw material, technology, and human resource potential as well as creating, through the energy sector, a lever for the development of the economy to foster equitable transformation [119]. Current targets consist of increasing the share of its renewables in fuel and technology mix to 23% by 2030, including up to 32% net in the power sector, reducing the share of coal in electricity generation to 56%, and reducing GHG emissions by 30% (relative to 1990), as well as implementing nuclear power by 2050 [120]. Poland's energy policies, over the past two decades, underline an important instance of how a country's ability to change and diversify its energy mix, utilizing European Union standards, has helped it develop its leadership role in Central and Eastern Europe. In comparison to the case research of energy poverty in Ukraine, it is evident that being a member state of the European Union has geopolitical weight that cannot be understated. Geopolitical variables outside of foreign policy and international political behavior include a country's border as well as its bordering countries, climate, demography, natural resources, topography, and scientific capability. In the case of Central and Eastern Europe, Ukraine and Poland illustrate two neighboring countries with two drastically different energy management systems. It is evident that the energy politics of these two countries showcase two different scenarios in terms of energy economics. In terms of the major cities in both countries, it equates to less urban energy output [103] and lower economic levels in Ukraine and more urban energy output [121,122] and higher economic levels in Poland.

4. Conclusions

The premise of this essay is to structure important standards in the light of the rural-to-urban transition and correlate them with urban energy needs. Three distinct themes were reviewed and inter-related to develop awareness for an urbanizing world that can otherwise appear to have a business-as-usual outlook. The urban population boom continues to pressure the energy dimension with heavily weighted impacts on LDRs. Moving forward, sustainable urban energy will need to be viable, healthy, and environmentally sound [24]. To achieve this, reduced "resource inputs and environmental impacts [will need to be decoupled with] economic growth from energy consumption" [24], e.g., utilizing a closed energy loop form of production and decarbonizing energy resources. Although fossil fuels continue to be the preferred method of energy for cities (and countries by extension), an increased understanding is emerging that sustainable energy forms can supplement them as an alternative. The implementation of renewables in city environments is quickly becoming "energetically imperative" [24] as we incorporate different energy transition processes and look towards a future where energy can be abundant and clean. Such processes will need to be "cost-effective, sustainable, and beneficial for development" [24]. Key to this transition will be the will to invest in renewables (i.e., solar, wind, hydro, tidal, geothermal, and biomass), efficient infrastructure, and smart eco-city designs. As mentioned in our introduction, the process of urbanization can be seen (at first glance) as a challenge, but upon closer examination we believe it to be an opportunity to manage efficient and effective energy designs that can make the future more livable and sustainable. The smart, urban energy transition envisions a new urbanism and city development that interplays between engineering, innovation, and the social sciences [32], with landscape architects and urban designers at the forefront. Our ability to build integrative, technology-based, and energy-friendly components within the physical city will be essential. Renewables side-by-side with green infrastructure will play a crucial role in the smart and sustainable city of the future [36]. Energy-conscious planning and design will be resolution-specific, from the makeup of an entire city to individual building design and microclimates [38]. The internal design of cities will allow us to specifically innovate technologies when they become available in best-case scenario-based development.

The technical transition of energy-friendly technologies is focused on understanding how the changes in the energy mix from non-renewable to renewable can be achieved. It is a step from an internal view and design of cities to a technical understanding of energy mix development and models that are not based on facility generation and a distribution network framed around storage systems and feedback loops. A technical transition will incorporate far-reaching concepts, such as smart grids with artificial intelligence that can distribute and store electricity as well as balance production and transmission volumes relative to consumption needs [52]. Ideally, they would operate internationally and alleviate some geopolitical barriers while, at the same time, strengthening partnerships that might otherwise not exist. Technical advancements within energy storage systems have mostly been found in MDRs and are based on a number of different types of battery and capacitor systems. To date, these systems cannot fulfil all energy storage needs [55,57] and should be seen as transitory. As part of this transition, the reduction in emitted pollutants from cities is paramount and backed up by a wide range of (urban) energy mitigation strategies. Urban energy mitigation incorporates legislation from all levels of government, i.e., municipal, national, and international. A number of urban energy functional initiatives for the reduction in GHG emissions from cities work in combination with urban planning and all other aspects discussed in terms of their internal design. By extension, all technical innovation should favor alternative energy technologies to improve and to be effectively implemented. As urban emissions are reduced, important changes environmentally, socially, and economically will become evident, e.g., the implementation of the 3R strategy to reduce, reuse, and recycle could mitigate a waste-to-energy model-based city economy that promotes sustainability and generationally friendly development.

Obviously, policy making will play, in large part, a structural role in which we will entrust leaders to make decisions based on sound scientific discovery and community accord. The geopolitical aspect, from this perspective, seems somewhat childish, since human health and well-being should be considered a given. Nonetheless, as nation states are still heavily reliant on fossil fuels with no permanent end in sight, especially within LDRs, energy management and an economic re-shift—as noted in the case of Central and Eastern Europe—become important hurdles for a common cause. In the case of Ukraine, its energy poverty is compounded by its strong geopolitical troubles. In Poland, since its admission into the European Union, it has become a viable example of energy advancement in Central and Eastern Europe which continues to strongly diversify its total energy mix. The essay's limited case research is used only to elucidate geopolitical complexity and country-specific individuality. Outside the purview of this work, it is fair to state that further case research from around the world is needed. The energy challenges presented in this essay attempt to piece together some of these universal issues that different countries and cities face at the geographical, technological advancement, and economic levels. As such, different nation states face a variety of energy concerns to meet their energy needs. The energy re-shift stressed in the essay veers towards the philosophical (and historical, to some degree), our ability to change and get along with one another. Energy, the life blood of modern cities, should be seen as an opportunity to unite, act responsibly and sustainably, and innovate towards a healthier humanity for a sounder human-energy relationship.

Author Contributions: Conceptualization, visualization, writing—original draft preparation, supervision, project administration, G.T.C.; Methodology, validation, formal analysis, investigation, resources, data curation, writing—review and editing, funding acquisition, G.T.C., A.R., F.B., E.C., A.G.G. and A.O.-J. All authors have read and agreed to the published version of the manuscript.

Funding: This research received no external funding.

Data Availability Statement: All study protocol, documents, and data pertaining to this study are available from the corresponding author on reasonable request.

Acknowledgments: The authors from the Faculty of Economics at the University of Gdansk, School of Arts at the University of Gloucestershire, and Department of Management at the International Humanitarian University are grateful to the respective institutions for supporting the research and for time allocated to work on the project. **Disclaimer.** The statements, opinions, and data contained in the work is solely those of the individual authors and may not reflect the opinion of the respected institutions they represent.

Conflicts of Interest: The authors declare no conflict of interest.

Appendix A

Table A1. Synthesized terms and combinative keyword search used for the methodology.

Concept	Search Term
Rural-to-urban transition	"urban" OR "urbanization" OR "urban development" OR "urban population" OR "urban migration" OR "urban sprawl" OR "urban growth" OR "rural-to-urban"
Smart city landscape design and energy innovation (internal urban strategies)	AND "smart city" OR "city development" OR "smart agenda" OR "energy sustainability" OR "energy system" OR "built environment" OR "sustainable city" OR "energy landscape" OR "landscape design" OR "energy planning" OR "green infrastructure" OR "green walls" OR "green roof" OR "greenery" OR "edible green infrastructure" OR "urban agriculture" OR "eco-cities" OR "efficient infrastructure" OR "carbon footprint"
Technical transition of energy-friendly technologies (urban energy mitigation from non-renewable to renewable)	AND "non-renewable energy" OR "renewable energy" OR "solar power" OR "wind energy" OR "hydroelectric energy" OR "tidal energy" OR "wave energy" OR "geothermal energy" OR "biomass energy" OR "power grid" OR "smart grid" OR "electricity storage" OR "energy balance" OR "energy production" OR "energy transition" OR "alternative fuel" OR "alternative energy" OR "energy technologies" OR "greenhouse gas emissions" OR "energy sector" OR "energy initiatives" OR "sustainable energy"
Geopolitical energy management and the economic re-shift (Central and Eastern Europe case research)	AND "geopolitical energy change" OR "energy market" OR "energy management" OR "energy price" OR "energy economics" OR "energy politics" OR "energy poverty" OR "energy blackouts" OR "energy shortages" OR "energy regulations" OR "energy reform" OR "energy consumption" OR "price of fossil fuels" OR "energy policy" OR "USA" OR "Russia" OR "China" OR "Germany" OR "Ukraine" OR "Poland" OR "Central Europe" OR "Eastern Europe" OR "European Union"

Appendix B

Table A2. Urban energy function initiatives for the reduction in GHG emissions from cities.

Urban Energy Function	Indicator	References
Land use limitation	Urbanization policy aiming at green areas establishment and water reservoirs	Cilliers et al. [123]; Di Leo et al. [124]; Kaur and Garg [125]; Vandermeulen et al. [126]
	Green buildings' design codes	BMZ [69]; Douglas [127]; UNFCCC [73]
	Densification with accompanied protected areas and restricted zones for settlement establishment	Ali and Al-Kodmany [128]; Lemonsu et al. [129]; Næss [130]
	Transit-oriented development	Chang and Murakami [131]; Mees [132]; Saif et al. [133]
	Participatory approach to new development of urban spaces	Batty et al. [134]; Ferreira et al. [135]; Jim and Shan [136]; Russo et al. [137]
Green buildings	Global codes and standards for building implementation	Akbari et al. [138]; C40 Cities et al. [72]; Farr [127]
	Zero energy through solar photovoltaics and micro wind turbines	Buonocore et al. [139]; Hayat et al. [140]; Mancebo [141]
	District heating systems based on biomass or municipal waste	Frick et al. [142]; Patuzzi et al. [143]; Pei-dong et al. [144]; Tutt and Olt [145]
	Solar hot water systems	Chu and Majumdar [146]; Hayat et al. [140]; Nelson [147]
	Heat recovery systems	BMZ [69]; C40 Cities et al. [72]; El-Hawary [55]
	Green roofs and water retention	Mancebo [141]; Soderlund and Newman [148]; Tanaka et al. [149]; Thornbush [150]
	Municipal auditing, support, and financial investment incentives	Alduwijaya et al. [151]; El-Hawary [55]

Table A2. *Cont.*

Urban Energy Function	Indicator	References
Sustainable urban mobility	Worldwide dissemination of the sustainable urban mobility planning process	Glazener and Khreis [3]; Malik et al. [152]
	Electrification of public transport	Abdul-Azeez and Ho [153]; Malik et al. [152]; Trahey et al. [154]
	Fuel switch to hydrogen	Blanco et al. [155]; Hordeski [156]; Saeedmanesh et al. [58]; Staffell et al. [157]
	Smart city logistics	BMZ [69]; Ceder [158]; Sajdak and Velazquez-Marti [159]; UNFCCC [73]
	Citizen behavior change into modal shift, shared mobility, and electrification	Batty et al. [134]; Borhan et al. [160]; Laurino and Grimaldi [161]; Lavadinho [162]; Morency et al. [163]; Suchanek et al. [71]
Decentralized energy supply	Smart grids locally governed	Aikhuele et al. [164]; Brinkerink et al. [165]; Chatzivasileiadis et al. [166]; El-Hawary [55]; Majeed Butt et al. [167]; Saidani Neffati et al. [53]
	Locally managed energy supply power plants	Aikhuele et al. [164]; El-Hawary [55]; InterContinental Energy [68]; Knight and Riggs [168]; Majeed Butt et al. [167]; Saidani Neffati et al. [53]; UNESCO [169];
Solid waste, water, and sewage management	3R strategy: reduce-reuse-recycle	Seto et al. [74]; UNFCCC [170]; Zamroni et al. [171]
	Waste-to-energy plant construction	Cunningham and Cunningham [172]; Krishnan et al. [173]; Laurent et al. [174]; Moore et al. [19]; UNFCCC [73]
	Improved recycling	Albores et al. [175]; Carbfix [78]; Daskal et al. [176]; Mancebo [177]; Miao et al. [178]; Ragnheidardottir et al. [79]
	Wastewater treatment facilities development	Farr [127]; Lee and Chang [179]; Vymazal [180]
	Landfill	Albores et al. [175]; Daskal et al. [176]; Davis et al. [181]; Peri et al. [182]; Santalla et al. [183]
	Education	Abel [184]; Mangizvo [185]; Moore et al. [19]; van Dijk [186]
Renewable energy sources	Municipally contracted energy purchase	Ackerman et al. [187]; Batty et al. [134]; Chyong [188]; Gielen et al. [189]; InterContinental Energy [68]; Karagiannis and Soldatos [190]
	Local energy storage systems	Aquila et al. [191]; Cotula [192]; Ghaffour et al. [193]; InterContinental Energy [68]; Laurent et al. [193]; UNESCO [194]; Whittinghill et al. [195,196]; Whittinghill and Rowe [197]
	Financial incentives and renewable energy sources' implementation obligations	Abdmouleh et al. [198]; Al-Kodmany [199]; Bointner et al. [200]; Borys and Śleszyński [201]; Delucchi and Jacobson [202]; Kobayashi and Ikaruga [203]; Köbbing et al. [204]; Qadir et al. [205]; Saeedmanesh et al. [58]; Stremke and Koh [37]

References

1. UN. *World Urbanization Prospects: The 2018 Revision*; United Nations, Department of Economic and Social Affairs: New York, NY, USA, 2019.
2. Rifkin, J. *The Risk of Too Much City*; Washington Post: Washington, DC, USA, 2006.
3. Glazener, A.; Khreis, H. Transforming Our Cities: Best Practices towards Clean Air and Active Transportation. *Curr. Environ. Heal. Rep.* **2019**, *6*, 22–37. [CrossRef]
4. Chokhachian, A.; Perini, K.; Giulini, S.; Auer, T. Urban performance and density: Generative study on interdependencies of urban form and environmental measures. *Sustain. Cities Soc.* **2020**, *53*, 101952. [CrossRef]
5. PRB. 2020 World Population Data Sheet. Available online: https://interactives.prb.org/2020-wpds (accessed on 26 June 2021).
6. Sassen, S. *The Global City*; Princeton University Press: New York, NY, USA, 1991.
7. Glaeser, E.L. *Triumph of the City: How Our Greatest Invention Makes Us Richer, Smarter, Greener, Healthier and Happier*; Penguin: London, UK, 2011.
8. Sigler, T. Book Review: Triumph of the City: How Our Greatest Invention Makes Us Richer, Smarter, Healthier, and Happier. *Urban Aff. Rev.* **2012**, *48*, 141–142. [CrossRef]
9. Reiss, D. Review: Triumph of the City: How Our Greatest Invention Makes Us Richer, Smarter, Greener, Healthier, and Happier. *Environ. Plan. A Econ. Sp.* **2013**, *45*, 749–750. [CrossRef]
10. Elmqvist, T.; Redman, C.L.; Barthel, S.; Costanza, R. History of urbanization and the missing ecology. In *Urbanization, Biodiversity and Ecosystem Services: Challenges and Opportunities: A Global Assessment*; Elmqvist, T., Ed.; Springer: Dordrecht, The Netherlands, 2013; pp. 13–30. ISBN 9789400770881.

11. Davis, K. The Origin and Growth of Urbanization in the World. *Am. J. Sociol.* **1955**, *60*, 429–437. [CrossRef]
12. Thompson, W.S. Population. *Am. J. Sociol.* **1929**, *34*, 959–975. [CrossRef]
13. Zelinsky, W. The Hypothesis of the Mobility Transition. *Geogr. Rev.* **1971**, *61*, 219. [CrossRef]
14. Jacobs, J. *Cities and the Wealth of Nations: Principles of Economic Life*; Random House: New York, NY, USA, 1984.
15. Gouldson, A.; Colenbrander, S.; Sudmant, A.; Godfrey, N.; Millward-Hopkins, J.; Fang, W.; Zhao, X. *Accelerating Low-Carbon Development in the World's Cities*; New Climate Economy: London, UK, 2015.
16. Bertinelli, L.; Strobl, E. Urbanisation, urban concentration and economic development. *Urban Stud.* **2007**, *44*, 2499–2510. [CrossRef]
17. Pumain, D. The urbanization process. In *Demography. Analysis and Synthesis*; Caselli, G., Vallin, J., Wunsch, G., Eds.; Elsevier: London, UK, 2006; pp. 319–328.
18. Benassi, F.; Heins, F.; Tucci, E. Residential Migrations in Italian Metropolitan Local Labour Market Areas: Spatial Patterns and Age-Structure Effects. In *Moving in Town. Practices, Pathways, and Contexts of Intra-Urban Mobility from 1600 to the Present Day*; Canepari, E., Crisci, M., Eds.; Casalini Libri: Fiesole, Italy, 2019; pp. 165–180.
19. Moore, M.; Gould, P.; Keary, B.S. Global urbanization and impact on health. *Int. J. Hyg. Environ. Health* **2003**, *206*, 269–278. [CrossRef] [PubMed]
20. Newman, P. The environmental impact of cities. *Environ. Urban.* **2006**, *18*, 275–295. [CrossRef]
21. Droege, P. *Urban Energy Transition*; Elsevier: Oxford, UK, 2008.
22. Riffat, S.; Powell, R.; Aydin, D. Future cities and environmental sustainability. *Future Cities Environ.* **2016**, *2*, 1. [CrossRef]
23. Asarpota, K.; Nadin, V. Energy strategies, the Urban dimension, and spatial planning. *Energies* **2020**, *13*, 3642. [CrossRef]
24. Phillips, L.; Smith, P. Sustainable Urban Energy Is the Future. *UN Chron.* **2013**, *52*, 23–25. [CrossRef]
25. Meyer, W.B. *The Environmental Advantages of Cities: Countering Commonsense Antiurbanism*; The MIT Press: Cambridge, MA, USA, 2013.
26. Mind Tools Starbursting: Understanding New Ideas by Brainstorming Questions. Available online: https://www.mindtools.com/pages/article/newCT_91.htm (accessed on 14 July 2021).
27. Ritter, S.M.; Mostert, N.M. How to facilitate a brainstorming session: The effect of idea generation techniques and of group brainstorm after individual brainstorm. *Creat. Ind. J.* **2018**, *11*, 263–277. [CrossRef]
28. Cirella, G.T.; Bąk, M.; Kozlak, A.; Pawłowska, B.; Borkowski, P. Transport innovations for elderly people. *Res. Transp. Bus. Manag.* **2019**, *30*, 100381. [CrossRef]
29. Matarese, V. Using strategic, critical reading of research papers to teach scientific writing: The reading-research-writing continuum. In *Supporting Research Writing*; Elsevier: Amsterdam, The Netherlands, 2013; pp. 73–89.
30. Renear, A.H.; Palmer, C.L. Strategic Reading, Ontologies, and the Future of Scientific Publishing. *Science* **2009**, *325*, 828–832. [CrossRef]
31. Broto, V.C. Energy landscapes and urban trajectories towards sustainability. *Energy Policy* **2017**, *108*, 755–764. [CrossRef]
32. Haarstad, H.; Wathne, M.W. Are smart city projects catalyzing urban energy sustainability? *Energy Policy* **2019**, *129*, 918–925. [CrossRef]
33. Hoang, A.T.; Pham, V.V.; Nguyen, X.P. Integrating renewable sources into energy system for smart city as a sagacious strategy towards clean and sustainable process. *J. Clean. Prod.* **2021**, *305*, 127161. [CrossRef]
34. Abu-Rayash, A.; Dincer, I. Development and analysis of an integrated solar energy system for smart cities. *Sustain. Energy Technol. Assess.* **2021**, *46*, 101170. [CrossRef]
35. Mozuriunaite, S. The role of landscape design in smart cities. *Landsc. Archit. Art* **2018**, *13*, 49–55. [CrossRef]
36. Blaschke, T.; Biberacher, M.; Gadocha, S.; Schardinger, I. 'Energy landscapes': Meeting energy demands and human aspirations. *Biomass Bioenergy* **2013**, *55*, 3–16. [CrossRef]
37. Stremke, S.; Koh, J. Integration of Ecological and Thermodynamic Concepts in the Design of Sustainable Energy Landscapes. *Landsc. J.* **2011**, *30*, 194–213. [CrossRef]
38. Haque, M.T.; Tai, L.; Ham, D. *Landscape Design for Energy Efficiency*; Clemson University Digital Press: Clemson, SC, USA, 2004; ISBN 097415167X.
39. European Commission Green Infrastructure in the Energy Sector. 2014. Available online: https://ec.europa.eu/environment/nature/ecosystems/pdf/Green%20Infrastructure/GI_energy.pdf (accessed on 10 July 2021).
40. Ko, Y. Trees and vegetation for residential energy conservation: A critical review for evidence-based urban greening in North America. *Urban For. Urban Green.* **2018**, *34*, 318–335. [CrossRef]
41. Wang, Z.-H.; Zhao, X.; Yang, J.; Song, J. Cooling and energy saving potentials of shade trees and urban lawns in a desert city. *Appl. Energy* **2016**, *161*, 437–444. [CrossRef]
42. McPherson, E.G.; Simpson, J.R. Potential energy savings in buildings by an urban tree planting programme in California. *Urban For. Urban Green.* **2003**, *2*, 73–86. [CrossRef]
43. Yuan, S.; Rim, D. Cooling energy saving associated with exterior greenery systems for three US Department of Energy (DOE) standard reference buildings. *Build. Simul.* **2018**, *11*, 625–631. [CrossRef]
44. Coma, J.; Pérez, G.; de Gracia, A.; Burés, S.; Urrestarazu, M.; Cabeza, L.F. Vertical greenery systems for energy savings in buildings: A comparative study between green walls and green facades. *Build. Environ.* **2017**, *111*, 228–237. [CrossRef]
45. Tan, H.; Hao, X.; Long, P.; Xing, Q.; Lin, Y.; Hu, J. Building envelope integrated green plants for energy saving. *Energy Explor. Exploit.* **2020**, *38*, 222–234. [CrossRef]

46. Kong, F.; Sun, C.; Liu, F.; Yin, H.; Jiang, F.; Pu, Y.; Cavan, G.; Skelhorn, C.; Middel, A.; Dronova, I. Energy saving potential of fragmented green spaces due to their temperature regulating ecosystem services in the summer. *Appl. Energy* **2016**, *183*, 1428–1440. [CrossRef]

47. Zhao, M.; Kong, Z.; Escobedo, F.J.; Gao, J. Impacts of urban forests on offsetting carbon emissions from industrial energy use in Hangzhou, China. *J. Environ. Manag.* **2010**, *91*, 807–813. [CrossRef] [PubMed]

48. Morabito, P. Green Infrastructure Can Yield Multiple Benefits in an Environmentally-Friendly Way: Three Examples of Green Infrastructure in Action. Available online: https://www.eesi.org/articles/view/green-infrastructure-can-yield-multiple-benefits-in-an-environmentally-frie (accessed on 10 July 2021).

49. Ferla, G.; Caputo, P.; Colaninno, N.; Morello, E. Urban greenery management and energy planning: A GIS-based potential evaluation of pruning by-products for energy application for the city of Milan. *Renew. Energy* **2020**, *160*, 185–195. [CrossRef]

50. Neuman, M. Spatial Planning Leadership by Infrastructure: An American View. *Int. Plan. Stud.* **2009**, *14*, 201–217. [CrossRef]

51. Hodge, J.; Haltrecht, J. *BedZED Seven Years on The Impact of the UK's Best Known Eco-Village and Its Residents*; BioRegional: London, UK, 2009.

52. Kougias, I.; Taylor, N.; Kakoulaki, G.; Jäger-Waldau, A. The role of photovoltaics for the European Green Deal and the recovery plan. *Renew. Sustain. Energy Rev.* **2021**, *144*, 111017. [CrossRef]

53. Saidani Neffati, O.; Sengan, S.; Thangavelu, K.D.; Dilip Kumar, S.; Setiawan, R.; Elangovan, M.; Mani, D.; Velayutham, P. Migrating from traditional grid to smart grid in smart cities promoted in developing country. *Sustain. Energy Technol. Assess.* **2021**, *45*, 101125. [CrossRef]

54. Sioshansi, F.P. *Smart Grid: Integrating Renewable, Distributed and Efficient Energy*; Elsevier: Amsterdam, The Netherlands, 2012; ISBN 9780-12-38-6452-9.

55. El-Hawary, M.E. The smart grid: State-of-the-art and future trends. *Electr. Power Components Syst.* **2014**, *42*, 239–250. [CrossRef]

56. Staffell, I.; Rustomji, M. Maximising the value of electricity storage. *J. Energy Storage* **2016**, *8*, 212–225. [CrossRef]

57. Abdelwahab, H.; Moussaid, L.; Moutaouakkil, F.; Medromi, H. Energy Efficiency: Improving the renewable energy penetration in a smart and green community. *Procedia Comput. Sci.* **2018**, *134*, 352–357. [CrossRef]

58. Saeedmanesh, A.; Mac Kinnon, M.A.; Brouwer, J. Hydrogen is essential for sustainability. *Curr. Opin. Electrochem.* **2018**, *12*, 166–181. [CrossRef]

59. Gul, M.; Kotak, Y.; Muneer, T. Review on recent trend of solar photovoltaic technology. *Energy Explor. Exploit.* **2016**, *34*, 485–526. [CrossRef]

60. BP. *Statistical Review of World Energy 2020*; BP: London, UK, 2020.

61. IRENA. *Innovation Landscape for a Renewable-Powered Future*; IRENA: Abu Dhabi, United Arab Emirates, 2019.

62. Deng, Y.Y.; Haigh, M.; Pouwels, W.; Ramaekers, L.; Brandsma, R.; Schimschar, S.; Grözinger, J.; de Jager, D. Quantifying a realistic, worldwide wind and solar electricity supply. *Glob. Environ. Chang.* **2015**, *31*, 239–252. [CrossRef]

63. Renoald, J.; Hemalatha, V.; Punitha, R.; Sasikala, M.; Sasikala, M.B.E. Solar Roadways-The Future Rebuilding Infrastructure and Economy. *Int. J. Electr. Electron. Res.* **2016**, *4*, 14–19.

64. Papadimitriou, C.N.; Psomopoulos, C.S.; Kehagia, F. A review on the latest trend of Solar Pavements in Urban Environment. *Energy Procedia* **2019**, *157*, 945–952. [CrossRef]

65. Needell, D.R.; Phelan, M.E.; Hartlove, J.T.; Atwater, H.A. Solar power windows: Connecting scientific advances to market signals. *Energy* **2021**, *219*, 119567. [CrossRef]

66. Ulavi, T.; Hebrink, T.; Davidson, J.H. Analysis of a Hybrid Solar Window for Building Integration. *Energy Procedia* **2014**, *57*, 1941–1950. [CrossRef]

67. Fernández-González, R.; Suárez-García, A.; Feijoo, M.Á.Á.; Arce, E.; Díez-Mediavilla, M. Spanish Photovoltaic Solar Energy: Institutional Change, Financial Effects, and the Business Sector. *Sustainability* **2020**, *12*, 1892. [CrossRef]

68. InterContinental Energy. Western Green Energy Hub. Available online: https://intercontinentalenergy.com/western-green-energy-hub (accessed on 23 July 2021).

69. BMZ. *Climate Change Mitigation in Cities: Urban Action to Reduce Greenhouse Gas Emissions*; Federal Ministry for Economic Cooperation and Development: Berlin, Germany, 2021.

70. World Resources Institute; C40 Cities Climate Leadership Group; ICLEI—Local Governments for Sustainability. *GHG Protocol for Cities*; Greenhouse Gas Protocol: Washington, DC, USA, 2014.

71. Suchanek, M.; Jagiełło, A.; Wołek, M. Transport Behaviour in the Context of Shared Mobility. In *Challenges of Urban Mobility, Transport Companies and Systems*; Suchanek, M., Ed.; Springer: Cham, Switzerland, 2018; pp. 149–158.

72. C40 Cities; Arup; University of Leeds. *The Future of Urban Consumption in a 1.5 °C World C40 Cities: Headline Report*; C40 Cities: Washington, DC, USA, 2019.

73. UNFCCC. *Urban Environment Related Mitigation Benefits and Co-Benefits of Policies, Practices and Actions for Enhancing Mitigation Ambition and Options for Supporting Their Implementation*; United Nations Framework Convention on Climate Change: Bonn, Germany, 2020.

74. Seto, K.C.; Bigio, A.; Blanco, H.; Carlo Delgado, G.; Bento, A.; Betsill, M.; Bulkeley, H.; Chavez, A.; Cervero, R.; Torres Martinez, J.; et al. Human Settlements, Infrastructure, and Spatial Planning. In *Climate Change 2014: Mitigation of Climate Change. Contribution of Working Group III to the Fifth Assessment Report of the Intergovernmental Panel on Climate Change*; Edenhofer, O., Pichs-Madruga, R., Sokona, Y., Farahani, E., Kadner, S., Seyboth, K., Adler, A., Baum, I., Brunner, S., Eickemeier, P., et al., Eds.; Cambridge University Press: Cambridge, UK, 2014.
75. Santamouris, M.; Paolini, R.; Haddad, S.; Synnefa, A.; Garshasbi, S.; Hatvani-Kovacs, G.; Gobakis, K.; Yenneti, K.; Vasilakopoulou, K.; Feng, J.; et al. Heat mitigation technologies can improve sustainability in cities. An holistic experimental and numerical impact assessment of urban overheating and related heat mitigation strategies on energy consumption, indoor comfort, vulnerability and heat-related mortality and morbidity in cities. *Energy Build.* **2020**, *217*, 110002. [CrossRef]
76. Santamouris, M.; Haddad, S.; Saliari, M.; Vasilakopoulou, K.; Synnefa, A.; Paolini, R.; Ulpiani, G.; Garshasbi, S.; Fiorito, F. On the energy impact of urban heat island in Sydney: Climate and energy potential of mitigation technologies. *Energy Build.* **2018**, *166*, 154–164. [CrossRef]
77. Zawadzka, J.E.; Harris, J.A.; Corstanje, R. Assessment of heat mitigation capacity of urban greenspaces with the use of InVEST urban cooling model, verified with day-time land surface temperature data. *Landsc. Urban Plan.* **2021**, *214*, 104163. [CrossRef]
78. Carbfix We Turn CO2 Into Stone. Available online: https://www.carbfix.com (accessed on 26 August 2021).
79. Ragnheidardottir, E.; Sigurdardottir, H.; Kristjansdottir, H.; Harvey, W. Opportunities and challenges for CarbFix: An evaluation of capacities and costs for the pilot scale mineralization sequestration project at Hellisheidi, Iceland and beyond. *Int. J. Greenh. Gas Control* **2011**, *5*, 1065–1072. [CrossRef]
80. City of Vancouver Climate Emergency Action Plan. Available online: https://vancouver.ca/green-vancouver/vancouvers-climate-emergency.aspx (accessed on 14 July 2021).
81. UN Goal 11 | Department of Economic and Social Affairs. Available online: https://sdgs.un.org/goals/goal11 (accessed on 14 July 2021).
82. Non-State Action Zone for Climate Action UNFCCC. Available online: https://unfccc.int (accessed on 5 July 2021).
83. European Commission. Covenant of Mayors for Climate and Energy: Europe. Available online: https://www.covenantofmayors.eu (accessed on 5 July 2021).
84. Ipek, P. Oil and intra-state conflict in Iraq and Syria: Sub-state actors and challenges for Turkey's energy security. *Middle East. Stud.* **2017**, *53*, 406–419. [CrossRef]
85. Hurewitz, J.C. *Oil, the Arab-Israel Dispute, and the Industrial World: Horizons of Crisis*; Routledge: New York, NY, USA, 2019; ISBN 9780367170691.
86. Rosales, A.; Sánchez, M. The Energy Politics of Venezuela. In *The Oxford Handbook of Energy Politics*; Hancock, K.J., Allison, J.E., Eds.; Oxford University Press: Oxford, UK, 2021; pp. 643–662.
87. Herberg, M.E. Asia's Growing Hunger for Energy: U.S. Policy and Supply Opportunities. Available online: https://www.nbr.org/publication/asias-growing-hunger-for-energy-u-s-policy-and-supply-opportunities (accessed on 24 June 2021).
88. Wang, D. The "Belt and Road" and the Safety of Maritime Energy Transportation Channels. In *China-Gulf Oil Cooperation under the Belt and Road Initiative*; Springer: Singapore, 2021; pp. 67–108.
89. BBC. Egypt's Suez Canal Blocked by Huge Container Ship. Available online: https://www.bbc.com/news/world-middle-east-56505413 (accessed on 24 June 2021).
90. Tabatabai, A.; Esfandiary, D. *Triple Axis: Iran's Relations with Russia and China*; Bloomsbury Publishing: London, UK, 2020; ISBN 9781788312394.
91. Bolt, P.J.; Cross, S.N. *China, Russia, and Twenty-First Century Global Geopolitics*; Oxford University Press: Oxford, UK, 2018; ISBN 9780198719519.
92. Zhou, Q.; He, Z.; Yang, Y. Energy geopolitics in Central Asia: China's involvement and responses. *J. Geogr. Sci.* **2020**, *30*, 1871–1895. [CrossRef]
93. Zhao, Y.; Liu, X.; Wang, S.; Ge, Y. Energy relations between China and the countries along the Belt and Road: An analysis of the distribution of energy resources and interdependence relationships. *Renew. Sustain. Energy Rev.* **2019**, *107*, 133–144. [CrossRef]
94. Mostafaeipour, A.; Alvandimanesh, M.; Najafi, F.; Issakhov, A. Identifying challenges and barriers for development of solar energy by using fuzzy best-worst method: A case study. *Energy* **2021**, *226*, 120355. [CrossRef]
95. Alsharif, M.H.; Kim, J.; Kim, J.H. Opportunities and challenges of solar and wind energy in South Korea: A review. *Sustainability* **2018**, *10*, 1822. [CrossRef]
96. Sourgens, F.G. The Biden (Energy) Doctrine. *ILSA J. Int. Comp. Law* **2021**, *27*, 293–314.
97. BBC. Nord Stream 2: Biden Waives US Sanctions on Russian Pipeline. Available online: https://www.bbc.com/news/world-us-canada-57180674 (accessed on 24 June 2021).
98. Nord Stream 2. Nord Stream 2 Construction. Available online: https://www.nord-stream2.com (accessed on 15 July 2021).
99. Goncharuk, A.G.; lo Storto, C. Challenges and policy implications of gas reform in Italy and Ukraine: Evidence from a benchmarking analysis. *Energy Policy* **2017**, *101*, 456–466. [CrossRef]
100. Sziklai, B.R.; Kóczy, L.; Csercsik, D. The impact of Nord Stream 2 on the European gas market bargaining positions. *Energy Policy* **2020**, *144*, 111692. [CrossRef]
101. Karpinska, L.; Śmiech, S. Invisible energy poverty? Analysing housing costs in Central and Eastern Europe. *Energy Res. Soc. Sci.* **2020**, *70*, 101670. [CrossRef]

102. Goncharuk, A.G.; Cirella, G.T. A perspective on household natural gas consumption in Ukraine. *Extr. Ind. Soc.* **2020**, *7*, 587–592. [CrossRef]
103. Goncharuk, A.G.; Hromovenko, K.V.; Pahlevanzade, A.; Hrinchenko, Y. Energy poverty leap during the pandemic: A case of Ukraine. *Polityka Energ.* **2021**, *24*, 5–18. [CrossRef]
104. Bouzarovski, S.; Tirado Herrero, S. The energy divide: Integrating energy transitions, regional inequalities and poverty trends in the European Union. *Eur. Urban Reg. Stud.* **2017**, *24*, 69–86. [CrossRef] [PubMed]
105. Bouzarovski, S.; Thomson, H.; Cornelis, M. Confronting Energy Poverty in Europe: A Research and Policy Agenda. *Energies* **2021**, *14*, 858. [CrossRef]
106. Leszczyńska, A.; Ki-Hoon, L. Sources and barriers of energy efficiency of Polish enterprises. *Ann. Univ. Mariae Curie-Skłodowska Lublin-Polonia* **2016**, *3*, 105–111. [CrossRef]
107. European Parliament. *Directive 2006/32/EC of the European Parliament and of the Council of 5 April 2006 on Energy End-Use Efficiency and Energy Services and Repealing Council Directive 93/76/EEC*; European Parliament: Brussels, Belgium, 2006.
108. Krawczyk, J.M.; Suwała, W. Directions for improving energy efficiency in Poland. *Energy Policy J.* **2014**, *17*, 226–231.
109. Government of Poland. *Law of April 15 2011 on Energy Efficiency*; Government of Poland: Warsaw, Poland, 2011.
110. Malon Group Wdrażanie. ISO 50001—Zarządzanie Energią Obniżające Koszty Zużycia Energii. Available online: https://www.iso.org.pl/uslugi-zarzadzania/wdrazanie-systemow/zarzadzanie-srodowiskowe/iso-50001 (accessed on 24 June 2021).
111. European Parliament. *Directive 2012/27/EU of the European Parliament and of the Council of 25 October 2012 on Energy Efficiency, Amending Directives 2009/125/EC and 2010/30/EU and Repealing Directives 2004/8/EC and 2006/32/EC Text with EEA Relevance*; European Parliament: Brussels, Belgium, 2012.
112. European Parliament. *Directive (EU) 2018/2002 of the European Parliament and of the Council of 11 December 2018 Amending Directive 2012/27/EU on Energy Efficiency*; European Parliament: Brussels, Belgium, 2018.
113. Government of Poland. *Law of July 19 2019 on the Compensation System for Energy-Intensive Sectors and Subsectors*; Government of Poland: Warsaw, Poland, 2019.
114. Energy Regulatory Office. *National Report: President of the Energy Regulatory Office*; Energy Regulatory Office: Warsaw, Poland, 2020.
115. Peryt, S.; Wnuk, R.; Berent-Kowalska, G.; Nowakowski, P. *Energy Efficiency in Poland in Years 2008–2018, Statistical Analyses*; Statistics Poland, Polish National Energy Conservation Agency: Warsaw, Poland, 2020.
116. Szczepaniak, K. Energy management system under conditions of sustainable development. *J. Manag. Financ.* **2014**, *12*, 390–401.
117. Komorowska, A.; Mirowski, T. Instruments for improving energy efficiency in Poland. *Zesz. Nauk. Inst. Gospod. Surowcami Miner. Energią Pol. Akad. Nauk.* **2016**, *92*, 300–314.
118. Jovane, F.; Yoshikawa, H.; Alting, L.; Boër, C.R.; Westkamper, E.; Williams, D.; Tseng, M.; Seliger, G.; Paci, A.M. The incoming global technological and industrial revolution towards competitive sustainable manufacturing. *CIRP Ann.—Manuf. Technol.* **2008**, *57*, 641–659. [CrossRef]
119. Monitor Polski. *Announcement of the Minister of Climate and Environment of 2 March 2021 on the National Energy Policy until 2040*; Government of Poland: Warsaw, Poland, 2021.
120. Government of Poland. *Sector Analysis: Branch Report*; Government of Poland: Warsaw, Poland, 2020.
121. Energy Cities Bielsko-Biala: Where Poland Shows that "Low-Carbon" Is Possible. Available online: https://energy-cities.eu/bielsko-biala-where-poland-shows-that-low-carbon-is-possible (accessed on 27 August 2021).
122. EIB Poland: The City of Chrzanów Will Build an Eco-District with the Support of the European Investment Advisory Hub. Available online: https://www.eib.org/en/press/all/2021-224-poland-the-city-of-chrzanow-will-build-an-eco-district-with-the-support-of-the-european-investment-advisory-hub (accessed on 26 August 2021).
123. Cilliers, S.; Cilliers, J.; Lubbe, R.; Siebert, S. Ecosystem services of urban green spaces in African countries—Perspectives and challenges. *Urban Ecosyst.* **2013**, *16*, 681–702. [CrossRef]
124. Di Leo, N.; Escobedo, F.J.; Dubbeling, M. The role of urban green infrastructure in mitigating land surface temperature in Bobo-Dioulasso, Burkina Faso. *Environ. Dev. Sustain.* **2016**, *18*, 373–392. [CrossRef]
125. Kaur, H.; Garg, P. Urban sustainability assessment tools: A review. *J. Clean. Prod.* **2019**, *210*, 146–158. [CrossRef]
126. Vandermeulen, V.; Verspecht, A.; Vermeire, B.; Van Huylenbroeck, G.; Gellynck, X. The use of economic valuation to create public support for green infrastructure investments in urban areas. *Landsc. Urban Plan.* **2011**, *103*, 198–206. [CrossRef]
127. Farr, D. *Sustainable Urbanism: Urban Design with Nature*; Wiley: New Jersey, NJ, USA, 2012; ISBN 1118174518.
128. Ali, M.M.; Al-Kodmany, K. Tall Buildings and Urban Habitat of the 21st Century: A Global Perspective. *Buildings* **2012**, *2*, 384–423. [CrossRef]
129. Lemonsu, A.; Viguié, V.; Daniel, M.; Masson, V. Vulnerability to heat waves: Impact of urban expansion scenarios on urban heat island and heat stress in Paris (France). *Urban Clim.* **2015**, *14*, 586–605. [CrossRef]
130. Næss, P. Urban Form, Sustainability and Health: The Case of Greater Oslo. *Eur. Plan. Stud.* **2014**, *22*, 1524–1543. [CrossRef]
131. Chang, Z.; Murakami, J. Transferring land-use rights with transportation infrastructure extensions: Evidence on spatiotemporal price formation in Shanghai. *J. Transp. Land Use* **2019**, *12*, 1–19. [CrossRef]
132. Mees, P. TOD and Multi-modal Public Transport. *Plan. Pract. Res.* **2014**, *29*, 461–470. [CrossRef]
133. Saif, M.A.; Zefreh, M.M.; Torok, A. Public transport accessibility: A literature review. *Period. Polytech. Transp. Eng.* **2019**, *47*, 36–43. [CrossRef]

134. Batty, M.; Axhausen, K.W.; Giannotti, F.; Pozdnoukhov, A.; Bazzani, A.; Wachowicz, M.; Ouzounis, G.; Portugali, Y. Smart cities of the future. *Eur. Phys. J. Spec. Top.* **2012**, *214*, 481–518. [CrossRef]
135. Ferreira, A.J.D.; Guilherme, R.I.M.M.; Ferreira, C.S.S.; Oliveira, M.D.F.M.L. Urban agriculture, a tool towards more resilient urban communities? *Curr. Opin. Environ. Sci. Health* **2018**, *5*, 93–97. [CrossRef]
136. Jim, C.Y.; Shan, X. Socioeconomic effect on perception of urban green spaces in Guangzhou, China. *Cities* **2013**, *31*, 123–131. [CrossRef]
137. Russo, A.; Chan, W.T.; Cirella, G.T. Estimating Air Pollution Removal and Monetary Value for Urban Green Infrastructure Strategies Using Web-Based Applications. *Land* **2021**, *10*, 788. [CrossRef]
138. Akbari, H.; Pomerantz, M.; Taha, H. Cool surfaces and shade trees to reduce energy use and improve air quality in urban areas. *Sol. Energy* **2001**, *70*, 295–310. [CrossRef]
139. Buonocore, J.J.; Hughes, E.J.; Michanowicz, D.R.; Heo, J.; Allen, J.G.; Williams, A. Climate and health benefits of increasing renewable energy deployment in the United States. *Environ. Res. Lett.* **2019**, *14*, 114010. [CrossRef]
140. Hayat, M.B.; Ali, D.; Monyake, K.C.; Alagha, L.; Ahmed, N. Solar energy-A look into power generation, challenges, and a solar-powered future. *Int. J. Energy Res.* **2019**, *43*, 1049–1067. [CrossRef]
141. Mancebo, F. Urban Agriculture for Urban Regeneration in the Sustainable City. In *Quality of Life in Urban Landscapes*; Grifoni, R.C., D'Onofrio, R., Sargolini, M., Eds.; Springer: Cham, Switzerland, 2018; pp. 311–317.
142. Frick, A.; Steffenhagen, P.; Zerbe, S.; Timmermann, T.; Schulz, K. Monitoring of the vegetation composition in rewetted peatland with iterative decision tree classification of satellite imagery. *Photogramm. Fernerkund. Geoinf.* **2011**, *2011*, 109–122. [CrossRef]
143. Patuzzi, F.; Mimmo, T.; Cesco, S.; Gasparella, A.; Baratieri, M. Common reeds (*Phragmites australis*) as sustainable energy source: Experimental and modelling analysis of torrefaction and pyrolysis processes. *GCB Bioenergy* **2013**, *5*, 367–374. [CrossRef]
144. Pei-dong, Z.; Guomei, J.; Gang, W. Contribution to emission reduction of CO_2 and SO_2 by household biogas construction in rural China. *Renew. Sustain. Energy Rev.* **2007**, *11*, 1903–1912. [CrossRef]
145. Tutt, M.; Olt, J. Suitability of various plant species for bioethanol production. *Agron. Res.* **2011**, *9*, 261–267.
146. Chu, S.; Majumdar, A. Opportunities and challenges for a sustainable energy future. *Nature* **2012**, *488*, 294–303. [CrossRef] [PubMed]
147. Nelson, V. *Introduction to Renewable Energy*; Ghassemi, A., Ed.; CRC Press: Boca Raton, FL, USA, 2011; ISBN 1439834504.
148. Soderlund, J.; Newman, P. Biophilic architecture: A review of the rationale and outcomes. *AIMS Environ. Sci.* **2015**, *2*, 950–969. [CrossRef]
149. Tanaka, Y.; Kawashima, S.; Hama, T.; Sánchez Sastre, L.F.; Nakamura, K.; Okumoto, Y. Mitigation of heating of an urban building rooftop during hot summer by a hydroponic rice system. *Build. Environ.* **2016**, *96*, 217–227. [CrossRef]
150. Thornbush, M. Urban agriculture in the transition to low carbon cities through urban greening. *AIMS Environ. Sci.* **2015**, *2*, 852–867. [CrossRef]
151. Ardiwijaya, V.S.; Sumardi, T.P.; Suganda, E.; Temenggung, Y.A. Rejuvenating Idle Land to Sustainable Urban form: Case Study of Bandung Metropolitan Area, Indonesia. *Procedia Environ. Sci.* **2015**, *28*, 176–184. [CrossRef]
152. Malik, Y.; Prakash, N.; Kapoor, A. Green transport: A way forward for environmental sustainability. In *Research in Political Sociology*; Emerald Group Publishing Ltd.: London, UK, 2018; Volume 25, pp. 163–180. ISBN 978-1-78714-776-8.
153. Abdul-Azeez, I.A.; Ho, C.S. Realizing Low Carbon Emission in the University Campus towards Energy Sustainability. *Open J. Energy Effic.* **2015**, *4*, 15–27. [CrossRef]
154. Trahey, L.; Brushett, F.R.; Balsara, N.P.; Ceder, G.; Cheng, L.; Chiang, Y.M.; Hahn, N.T.; Ingram, B.J.; Minteer, S.D.; Moore, J.S.; et al. Energy storage emerging: A perspective from the Joint Center for Energy Storage Research. *Proc. Natl. Acad. Sci. USA* **2020**, *117*, 12550–12557. [CrossRef]
155. Blanco, H.; Nijs, W.; Ruf, J.; Faaij, A. Potential for hydrogen and Power-to-Liquid in a low-carbon EU energy system using cost optimization. *Appl. Energy* **2018**, *232*, 617–639. [CrossRef]
156. Hordeski, M.F. *Alternative Fuels: The Future of Hydrogen*; Fairmont Press: Lilburn, GA, USA, 2008; ISBN 0881735965.
157. Staffell, I.; Scamman, D.; Velazquez Abad, A.; Balcombe, P.; Dodds, P.E.; Ekins, P.; Shah, N.; Ward, K.R. The role of hydrogen and fuel cells in the global energy system. *Energy Environ. Sci.* **2019**, *12*, 463–491. [CrossRef]
158. Ceder, A. Urban mobility and public transport: Future perspectives and review. *Int. J. Urban Sci.* **2020**, 1–25. [CrossRef]
159. Sajdak, M.; Velazquez-Marti, B. Estimation of pruned biomass form dendrometric parameters on urban forests: Case study of Sophora japonica. *Renew. Energy* **2012**, *47*, 188–193. [CrossRef]
160. Borhan, M.N.; Syamsunur, D.; Mohd Akhir, N.; Mat Yazid, M.R.; Ismail, A.; Rahmat, R.A. Predicting the use of public transportation: A case study from Putrajaya, Malaysia. *Sci. World J.* **2014**, *2014*, 1–9. [CrossRef]
161. Laurino, A.; Grimaldi, R. The Italian Way to Carsharing. *TeMA J. Land Use Mobil. Environ.* **2012**, *5*, 77–90. [CrossRef]
162. Lavadinho, S. Public transport infrastructure and walking: Gearing towards the multimodal city. In *Transport and Sustainability*; Emerald Group Publishing Ltd.: London, UK, 2017; Volume 9, pp. 167–186. ISBN 978-1-78714-628-0.
163. Morency, C.; Habib, K.M.N.; Grasset, V.; Islam, M.T. Understanding members' carsharing (activity) persistency by using econometric model. *J. Adv. Transp.* **2012**, *46*, 26–38. [CrossRef]
164. Aikhuele, D.O.; Ighravwe, D.E.; Akinyele, D. Evaluation of Renewable Energy Technology Based on Reliability Attributes Using Hybrid Fuzzy Dynamic Decision-Making Model. *Technol. Econ. Smart Grids Sustain. Energy* **2019**, *4*, 1–7. [CrossRef]

165. Brinkerink, M.; Gallachóir, B.; Deane, P. A comprehensive review on the benefits and challenges of global power grids and intercontinental interconnectors. *Renew. Sustain. Energy Rev.* **2019**, *107*, 274–287. [CrossRef]
166. Chatzivasileiadis, S.; Ernst, D.; Andersson, G. The Global Grid. *Renew. Energy* **2013**, *57*, 372–383. [CrossRef]
167. Majeed Butt, O.; Zulqarnain, M.; Majeed Butt, T. Recent advancement in smart grid technology: Future prospects in the electrical power network. *Ain Shams Eng. J.* **2020**, in press. [CrossRef]
168. Knight, L.; Riggs, W. Nourishing urbanism: A case for a new urban paradigm. *Int. J. Agric. Sustain.* **2010**, *8*, 116–126. [CrossRef]
169. UNESCO. The United Nations World Water Development Report 3: Water in a Changing World. In *World Water Assessment Programme*; Earthscan, Ed.; UNESCO: Paris, France, 2009; Volume 3, p. 349. ISBN 9789231042355.
170. UNFCCC. United Nations: Climate Change. Available online: https://unfccc.int (accessed on 14 February 2021).
171. Zamroni, M.; Prahara, R.S.; Kartiko, A.; Purnawati, D.; Kusuma, D.W. The Waste Management Program Of 3R (Reduce, Reuse, Recycle) By Economic Incentive and Facility Support. *J. Phys. Conf. Ser.* **2020**, *1471*, 12048. [CrossRef]
172. Cunningham, W.P.; Cunningham, M.A. *Principles of Environmental Science: Inquiry and Applications*; McGraw-Hill: New York, NY, USA, 2006; ISBN 0073019267.
173. Krishnan, R.; Geyskens, I.; Steenkamp, J.B.E.M. The effectiveness of contractual and trust-based governance in strategic alliances under behavioral and environmental uncertainty. *Strateg. Manag. J.* **2016**, *37*, 2521–2542. [CrossRef]
174. Laurent, A.; Bakas, I.; Clavreul, J.; Bernstad, A.; Niero, M.; Gentil, E.; Hauschild, M.Z.; Christensen, T.H. Review of LCA studies of solid waste management systems—Part I: Lessons learned and perspectives. *Waste Manag.* **2014**, *34*, 573–588. [CrossRef]
175. Albores, P.; Petridis, K.; Dey, P.K. Analysing Efficiency of Waste to Energy Systems: Using Data Envelopment Analysis in Municipal Solid Waste Management. *Procedia Environ. Sci.* **2016**, *35*, 265–278. [CrossRef]
176. Daskal, S.; Ayalon, O.; Shechter, M. The state of municipal solid waste management in Israel. *Waste Manag. Res.* **2018**, *36*, 527–534. [CrossRef] [PubMed]
177. Mancebo, F. Gardening the City: Addressing Sustainability and Adapting to Global Warming through Urban Agriculture. *Environments* **2018**, *5*, 38. [CrossRef]
178. Miao, C.; Fang, D.; Sun, L.; Luo, Q. Natural resources utilization efficiency under the influence of green technological innovation. *Resour. Conserv. Recycl.* **2017**, *126*, 153–161. [CrossRef]
179. Lee, C.-S.; Chang, S.-P. Interactive fuzzy optimization for an economic and environmental balance in a river system. *Water Res.* **2005**, *39*, 221–231. [CrossRef]
180. Vymazal, J. Constructed wetlands for wastewater treatment. *Water* **2010**, *2*, 530–549. [CrossRef]
181. Davis, G.; Phillips, P.S.; Read, A.D.; Iida, Y. Demonstrating the need for the development of internal research capacity: Understanding recycling participation using the Theory of Planned Behaviour in West Oxfordshire, UK. *Resour. Conserv. Recycl.* **2006**, *46*, 115–127. [CrossRef]
182. Peri, G.; Traverso, M.; Finkbeiner, M.; Rizzo, G. The cost of green roofs disposal in a life cycle perspective: Covering the gap. *Energy* **2012**, *48*, 406–414. [CrossRef]
183. Santalla, E.; Córdoba, V.; Blanco, G. Greenhouse gas emissions from the waste sector in Argentina in business-as-usual and mitigation scenarios. *J. Air Waste Manag. Assoc.* **2013**, *63*, 909–917. [CrossRef]
184. Abel, A. An analysis of solid waste generation in a traditional African city: The example of Ogbomoso, Nigeria. *Environ. Urban.* **2007**, *19*, 527–537. [CrossRef]
185. Mangizvo, R.V. Challenges of Solid Waste Management in the Central Business District of the City of Gweru in Zimbabwe. *J. Sustain. Dev.* **2007**, *9*, 134–145.
186. Van Dijk, M.P. Three Ecological Cities, Examples of Different Approaches in Asia and Europe. In *Eco-City Planning*; Wong, T.-C., Yuen, B., Eds.; Springer: Dordrecht, The Netherlands, 2011; pp. 31–50. ISBN 978-94-007-0383-4.
187. Ackerman, K.; Conard, M.; Culligan, P.; Plunz, R.; Sutto, M.P.; Whittinghill, L. Sustainable food systems for future cities: The potential of urban agriculture. *Econ. Soc. Rev.* **2014**, *45*, 189–206.
188. Chyong, C.K. European Natural Gas Markets: Taking Stock and Looking Forward. *Rev. Ind. Organ.* **2019**, *55*, 89–109. [CrossRef]
189. Gielen, D.; Boshell, F.; Saygin, D.; Bazilian, M.D.; Wagner, N.; Gorini, R. The role of renewable energy in the global energy transformation. *Energy Strateg. Rev.* **2019**, *24*, 38–50. [CrossRef]
190. Karagiannis, I.C.; Soldatos, P.G. Estimation of critical CO_2 values when planning the power source in water desalination: The case of the small Aegean islands. *Energy Policy* **2010**, *38*, 3891–3897. [CrossRef]
191. Aquila, G.; Pamplona, E.D.O.; Queiroz, A.R.D.; Rotela Junior, P.; Fonseca, M.N. An overview of incentive policies for the expansion of renewable energy generation in electricity power systems and the Brazilian experience. *Renew. Sustain. Energy Rev.* **2017**, *70*, 1090–1098. [CrossRef]
192. Cotula, L. The international political economy of the global land rush: A critical appraisal of trends, scale, geography and drivers. *J. Peasant Stud.* **2012**, *39*, 649–680. [CrossRef]
193. Ghaffour, N.; Missimer, T.M.; Amy, G.L. Technical review and evaluation of the economics of water desalination: Current and future challenges for better water supply sustainability. *Desalination* **2013**, *309*, 197–207. [CrossRef]
194. UNESCO. The United Nations World Water Development Report 4: Managing Water under Uncertainty and Risk. In *World Water Assessment Programme*; Earthscan, Ed.; UNESCO: Paris, France, 2012; Volume 1, p. 407. ISBN 9789231042355.
195. Whittinghill, L.J.; Rowe, D.B.; Schutzki, R.; Cregg, B.M. Quantifying carbon sequestration of various green roof and ornamental landscape systems. *Landsc. Urban Plan.* **2014**, *123*, 41–48. [CrossRef]

196. Whittinghill, L.J.; Rowe, D.B.; Andresen, J.A.; Cregg, B.M. Comparison of stormwater runoff from sedum, native prairie, and vegetable producing green roofs. *Urban Ecosyst.* **2014**, *18*, 13–29. [CrossRef]
197. Whittinghill, L.J.; Rowe, D.B. The role of green roof technology in urban agriculture. *Renew. Agric. Food Syst.* **2012**, *27*, 314–322. [CrossRef]
198. Abdmouleh, Z.; Alammari, R.A.M.; Gastli, A. Review of policies encouraging renewable energy integration & best practices. *Renew. Sustain. Energy Rev.* **2015**, *45*, 249–262. [CrossRef]
199. Al-Kodmany, K. Green towers and iconic design: Cases from three continents. *Archnet-IJAR Int. J. Archit. Res.* **2014**, *8*, 11–28. [CrossRef]
200. Bointner, R.; Pezzutto, S.; Grilli, G.; Sparber, W. Financing innovations for the renewable energy transition in Europe. *Energies* **2016**, *9*, 990. [CrossRef]
201. Borys, T.; Śleszyński, J. Ekorozwój jako zbiór zasad. In *Wskaźniki Ekorozwoju. Wydawnictwo Ekonomia i Środowisko*; Borys, T., Ed.; Foundation of Environmental and Resource Economists: Białystok, Poland, 1999; pp. 85–92.
202. Delucchi, M.A.; Jacobson, M.Z. Providing all global energy with wind, water, and solar power, Part II: Reliability, system and transmission costs, and policies. *Energy Policy* **2011**, *39*, 1170–1190. [CrossRef]
203. Kobayashi, T.; Ikaruga, S. Development of a smart city planning support tool using the cooperative method. *Front. Archit. Res.* **2015**, *4*, 277–284. [CrossRef]
204. Köbbing, J.F.; Thevs, N.; Zerbe, S. The utilisation of Reed (Phragmites australis)—A review. *Mires Peat* **2013**, *13*, 1–14.
205. Qadir, S.A.; Al-Motairi, H.; Tahir, F.; Al-Fagih, L. Incentives and strategies for financing the renewable energy transition: A review. *Energy Reports* **2021**, *7*, 3590–3606. [CrossRef]

MDPI

St. Alban-Anlage 66

4052 Basel

Switzerland

Tel. +41 61 683 77 34

Fax +41 61 302 89 18

www.mdpi.com

Energies Editorial Office

E-mail: energies@mdpi.com

www.mdpi.com/journal/energies